SIGNIFICANT CHANGES TO THE

INTERNATIONAL PLUMBING CODE®, INTERNATIONAL MECHANICAL CODE®, AND INTERNATIONAL FUEL GAS CODE®

2012 EDITION

DELMAR
CENGAGE Learning™

Australia • Brazil • Japan • Korea • Mexico • Singapore • Spain • United Kingdom • United States

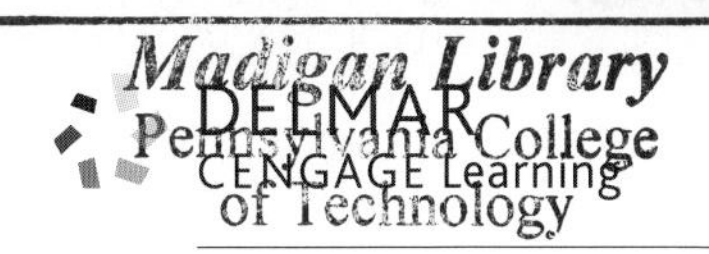

Significant Changes to the International Plumbing Code, International Mechanical Code, and International Fuel Gas Code 2012 Edition
International Code Council

Delmar Cengage Learning Staff:

Vice President, Technology and Trades Professional Business Unit: Gregory L. Clayton

Director of Building Trades: Taryn Zlatin McKenzie

Executive Editor: Robert Person

Development: Nobina Preston

Director of Marketing: Beth A. Lutz

Marketing Manager: Marissa Maiella

Marketing Communications Manager: Nicole McKasty Stagg

Senior Director, Education Production: Wendy A. Troeger

Production Director: Sherondra Thedford

Senior Content Project Manager: Stacey Lamodi

Senior Art Director: Benjamin Gleeksman

ICC Staff:

Senior Vice President, Business and Product Development: Mark A. Johnson

Deputy Senior Vice President, Business and Product Development: Hamid Naderi

Technical Director, Product Development: Doug Thornburg

Manager, Product and Special Sales: Suzane Nunes Holten

Senior Marketing Specialist: Dianna Hallmark

Library of Congress Control Number: 2011924617

ISBN-13: 978-1-111-54247-4

ISBN-10: 1-111-54247-3

ICC World Headquarters
500 New Jersey Avenue, NW
6th Floor
Washington, DC 20001-2070
Telephone: 1-888-ICC-SAFE (422-7233)
Website: **http://www.iccsafe.org**

Delmar
5 Maxwell Drive
Clifton Park, NY 12065-2919
USA

Cengage Learning is a leading provider of customized learning solutions with office locations around the globe, including Singapore, the United Kingdom, Australia, Mexico, Brazil, and Japan. Locate your local office at: **international.cengage.com/region**

Cengage Learning products are represented in Canada by Nelson Education, Ltd.

Visit us at **www.InformationDestination.com**
For more learning solutions, visit **www.cengage.com**

Notice to the Reader
Publisher does not warrant or guarantee any of the products described herein or perform any independent analysis in connection with any of the product information contained herein. Publisher does not assume, and expressly disclaims, any obligation to obtain and include information other than that provided to it by the manufacturer. The reader is expressly warned to consider and adopt all safety precautions that might be indicated by the activities described herein and to avoid all potential hazards. By following the instructions contained herein, the reader willingly assumes all risks in connection with such instructions. The publisher makes no representations or warranties of any kind, including but not limited to, the warranties of fitness for particular purpose or merchantability, nor are any such representations implied with respect to the material set forth herein, and the publisher takes no responsibility with respect to such material. The publisher shall not be liable for any special, consequential, or exemplary damages resulting, in whole or part, from the readers' use of, or reliance upon, this material.

Printed in the United States of America
1 2 3 4 5 6 7 14 13 12 11

Contents

Preface

The purpose of *Significant Changes to the International Plumbing Code, International Mechanical Code, and International Fuel Gas Code® 2012 Edition* is to familiarize plumbing and mechanical officials, building officials, fire officials, plans examiners, inspectors, design professionals, contractors, and others in the construction industry with many of the important changes in the 2012 *International Plumbing Code, International Mechanical Code*, and *International Fuel Gas Code* (IPC/IMC/IFGC). This publication is designed to assist code users in identifying the specific code changes that have occurred and, more important, in understanding the reasons behind the changes. It is also a valuable resource for jurisdictions in the code-adoption process.

Only portions of the total number of code changes to the IPC/IMC/IFGC are discussed in this book. The changes selected were identified for a number of reasons, including their frequency of application, special significance, or change in application. However, the importance of those changes not included is not to be diminished. Further information on all code changes can be found in the *Code Changes Resource Collection,* published by the International Code Council® (ICC®), which provides the published documentation for each successful code change contained in the 2012 IPC and 2009 IMC.

Throughout this significant changes book, each change is accompanied by a photograph, an application example, or an illustration to assist and enhance the reader's understanding of the specific change. A summary and discussion of the significance of the changes are also provided. Each code change is identified by type, be it an addition, modification, clarification, or deletion.

The code change itself is presented in a format similar to the style utilized for code-change proposals. Deleted code language is shown with a strikethrough, whereas new code text is indicated by underlining. As a result, the actual 2012 code language is provided, as well as a comparison with the 2009 language, so the user can easily determine changes to the specific code text.

As with any code-change text, *Significant Changes to the International Plumbing Code, International Mechanical Code, and International Fuel*

Gas Code 2012 Edition is best used as a study companion to the 2012 IPC, 2012 IMC, and 2012 IFGC. Because only a limited discussion of each change is provided, the code itself should always be referenced in order to gain a more comprehensive understanding of the code change and its application.

The commentary and opinions set forth in this text are those of the authors and do not necessarily represent the official position of the ICC. In addition, they may not represent the views of any enforcing agency, as such agencies have the sole authority to render interpretations of the code. In many cases, the explanatory material is derived from the reasoning expressed by code-change proponents.

Comments concerning this publication are encouraged and may be directed to the ICC at significantchanges@iccsafe.org.

About the *International Plumbing, International Mechanical,* and *International Fuel Gas Codes*

Code officials, design professionals, and others involved in the building construction industry recognize the need for a modern, up-to-date building code addressing the design and installation of building systems, including plumbing, mechanical, and fuel gas systems, through requirements emphasizing performance. The 2012 editions of the *International Plumbing Code®* (IPC), *International Mechanical Code®* (IMC), and *International Fuel Gas Code®* (IFGC) are intended to meet these needs through model code regulations that safeguard public health and safety in all communities, large and small. The IPC/IMC/IFGC are kept up to date through the ICC's open code-development process. The provisions of the 2009 editions, along with those code changes approved through 2010, make up the 2012 editions.

The ICC, publisher of the I-Codes, was established in 1994 as a nonprofit organization dedicated to developing, maintaining, and supporting a single set of comprehensive and coordinated national model building construction codes. Its mission is to provide the highest-quality codes, standards, products, and services for all concerned with the safety and performance of the built environment.

The IPC, IMC, and IFGC are three of the 13 International Codes® published by the ICC. These comprehensive codes establish minimum regulations for plumbing, mechanical, and fuel gas systems by means of prescriptive and performance-related provisions and are founded on broad-based principles that make possible the use of new materials and new system designs. The IPC, IMC, and IFGC are available for adoption and use by jurisdictions internationally. Their use within a governmental jurisdiction is intended to be accomplished through adoption by reference, in accordance with proceedings establishing the jurisdiction's laws.

Acknowledgments

Lee Clifton, author of the IPC section, thanks the PMG members of the ICC for their assistance in the preparation of this book.

Lee is grateful to his father, Bill Clifton, and his mother, Lila Lee Clifton, both of the Tampa, Florida family run plumbing business, William E. Clifton Plumbing Inc., where Lee began learning the plumbing trade and was blessed with encouragement and patience of his parents.

He thanks his wife, Mary Lou, for understanding and supporting his decision to retire from his 25 year principal plumbing inspector position with the City of Los Angeles to pursue his career with the ICC.

Bob Guenther, author of the IMC and IFGC portions of this book, would like to thank the ICC staff members that assisted with this publication, in particular Alexandria Pearce, Audrie Cetina, and Doug Thornburg.

About the Authors

Lee Clifton

Director of Plumbing Programs

International Code Council

Lee Clifton is the Director of Plumbing Programs for the International Code Council, Plumbing and Mechanical Activities. Prior to joining the Code Council, Lee was with the City of Los Angeles for 21 years. As the Principal Inspector with the City of Los Angeles, his duties included planning, organizing, and coordinating the training programs for inspectors and clerical staff. Lee has served on the Board of Directors for the Southern California Health and Housing Council and participated on the Strategic Planning Committee to end lead poisoning in Los Angeles County. In prior years he also served as a Senior Plumbing Training Officer for the Department of Building and Safety. Lee has taught the model plumbing code for the Plumbing-Heating-Cooling Contractors Association (PHCC) of the Greater Los Angeles Area and the North Orange County Regional Occupation Program. His career began as a second generation plumber, working for his father at Clifton Plumbing Inc. in Tampa, Florida, and then entering the United States Coast Guard. Lee completed his plumbing apprenticeship training in southern California.

Lee holds an L.A. City Journeyman Plumber's license; numerous certificates for training and medical gas inspection and an ICC Certified Plumbing and Mechanical Inspector. He is a member of the American Society of Sanitary Engineers, the American Society of Plumbing Engineers, and a member of the American Backflow Prevention Association. In the past Lee has participated on numerous industry committees and has authored many technical books and publications.

Bob Guenther
Director of Mechanical Programs
International Code Council

Bob Guenther is the Director of Mechanical Programs for the ICC in Whittier, California. He provides technical assistance to the members of the ICC on the IPC/IMC/IFGC. Bob also develops and reviews technical publications that involve mechanical, fuel gas, and plumbing material. Bob also presents mechanical and fuel gas code seminars throughout the country. Prior to coming to work for the ICC, Bob was the Mechanical Official for Long Beach, California, where he worked for 24 years, and prior to that he was a Mechanical Inspector for the City of Los Angeles for 5 years. Bob has taught the mechanical code in community colleges in southern California and has presented mechanical code seminars for 30 years. Bob has served on the IMC and Uniform Mechanical Code (UMC) code-changes committees and was on the committee that formed the original IMC.

About the International Code Council®

The International Code Council® (ICC®) is a nonprofit membership association dedicated to protecting the health, safety, and welfare of people by creating better buildings and safer communities. The mission of ICC is to provide the highest quality codes, standards, products and services for all concerned with the safety and performance of the built environment. ICC is the publisher of the family of the International Codes® (I-Codes®), a single set of comprehensive and coordinated model codes. This unified approach to building codes enhances safety, efficiency and affordability in the construction of buildings. The Code Council is also dedicated to innovation, sustainability and energy efficiency. Code Council subsidiary ICC Evaluation Service issues Evaluation Reports for innovative products and reports of Sustainable Attributes Verification and Evaluation (SAVE).

Headquarters:
500 New Jersey Avenue, NW, 6th Floor
Washington, DC 20001-2070

District Offices:
Birmingham, AL; Chicago. IL; Los Angeles, CA

1-888-422-7233
www.iccsafe.org

PART 1

International Plumbing Code

Chapters 1 through 14

- **Chapter 1** Administration No changes addressed
- **Chapter 2** Definitions
- **Chapter 3** General Regulations
- **Chapter 4** Fixture, Faucets, and Fixtures
- **Chapter 5** Water Heaters
- **Chapter 6** Water Supply Distribution
- **Chapter 7** Sanitary Drainage
- **Chapter 8** Indirect/Special Waste
- **Chapter 9** Vents
- **Chapter 10** Traps, Interceptors, and Separator
- **Chapter 11** Storm Drainage
- **Chapter 12** Special Piping and Storage No changes addressed
- **Chapter 13** Gray Water Recycling Systems
- **Chapter 14** Referenced Standards No changes addressed

Chapter 1 of the *International Plumbing Code*® (IPC) clarifies how the code will be enforced by code officials. Definitions of plumbing code terminology are found in Chapter 2. General regulations in Chapter 3 identify requirements not listed in other code chapters, such as testing and inspections. Fixtures and water heaters are addressed in Chapters 4 and 5, respectively. Chapters 6 and 7 regulate water and drainage piping systems, in that order. Indirect/special waste is covered in Chapter 8. Chapter 9 details acceptable venting methodologies with in-depth piping arrangements. The provisions for traps with various receptors are found in Chapter 10. Storm drainage, with its piping collection system, is covered in Chapter 11. Installation, design, storage, handling, and use of nonflammable medical gas systems are addressed in Chapter 12. Gray-water recycling systems are now addressed in Chapter 13. Standards are identified with clear guidelines in Chapter 14. Appendices A through F cover nonmandatory provisions for permit fees, rainfall rates, vacuum drainage, degree design temperatures, a water sizing method, and structural protection methodology. ■

202

Plumbing Fixture Definition

202

Plumbing Appliance Definition

202

Grease Interceptor Definition

303.1, 303.4

Material Identification and Third-Party Certification

308.9

Parallel Water Distribution Systems

315.1

Sealing of Annular Spaces at Penetrations

TABLE 403.1

Minimum Number of Required Plumbing Fixtures

403.2

Separate Toilet Facilities in Group M Occupancies

403.2.1

Family or Assisted-Use Toilet Facilities Serving as Separate Facilities

403.3.2

Relationship of Toilet Rooms and Food Preparation Areas

403.3.6

Locking of Toilet Room Doors

403.5

Drinking Fountain Locations

405.3.1

Minimum Water Closet Compartment Size

405.4

Floor and Wall Drainage Connections

407.2

Bathtub Waste Outlets and Overflows

410

Minimum Required Number of Drinking Fountains

417.5.2.6

Shower Pan Liner Materials

424.9

Water Closet Personal Hygiene Devices

504.4.1

Water Heater Storage Tank Relief Valves

504.7

Water Heater Pans

605

Polyethylene of Raised-Temperature (PE-RT) Plastic Tubing

TABLE 605.3

Polyethylene (PE) Water Service Pipe

TABLE 605.3

PEX Water Service Pipe

606.7

Labeling of Water Distribution Pipes in Bundles

607.1.1

Water-Temperature-Limiting Means

607.2

Hot or Tempered Water Supply to Fixtures

607.5

Hot Water Piping Insulation

608.8

Identification of Nonpotable Water

704.3, 711.2.1

Horizontal Branch Connections

TABLE 709.1

Drainage Fixture Units for Bathroom Groups

712.3.3

Sump Pump and Ejector Discharge Pipe and Fittings

712.3.5

Sump Pump Connection to the Drainage System

715.1

Fixture Protection from Sewage Backflow

802.1.8

Indirect Discharge of Food Preparation Sinks

802.2

Installation of Indirect Waste Piping

802.3

Prohibited Locations for Waste Receptors

901.3, 918.8

Air Admittance Valves for Chemical Waste Vent Systems

903.5

Location of Vent Terminals

915.2

Combination Waste and Vent System Sizing

917

Single-Stack Vent Systems

1002.1

Floor Drains in Multi-Level Parking Structures

1003.1

Interceptors and Separators

1003.3.1

Alternate Grease Interceptor Locations

1003.3.4

Hydromechanical Grease Interceptors

1105

Roof Drain Strainers

1107

Siphonic Roof Drainage Systems

CHAPTER 13

Gray-Water Recycling Systems

202

Plumbing Fixture Definition

International Code Council®

Waterless urinal

CHANGE TYPE: Modification

CHANGE SUMMARY: The definition of "plumbing fixture" has been modified to include fixtures such as waterless urinals.

2012 CODE: **Plumbing Fixture.** A receptacle or device that is ~~either permanently or temporarily~~ connected to ~~the~~ a water ~~distribution~~ supply system or ~~of the premises and demands a supply of water therefrom;~~ discharges ~~wastewater, liquid-borne waste materials or sewage either directly or indirectly~~ to a ~~the~~ drainage system ~~of the premises;~~ or ~~requires~~ both. Such receptacles or devices require a ~~water~~ supply of water; ~~connection and a~~ or discharge liquid waste or liquid-borne solid waste; or require a supply of water and discharge waste to a ~~the~~ drainage system ~~of the premises~~.

CHANGE SIGNIFICANCE: The previous definition of "plumbing fixture" was outdated and incomplete. The definition now includes receptacles and devices that do not necessarily require connection to a water supply. Waterless urinals and floor drains, now addressed in the revised definition, were not defined as plumbing fixtures in the past.

202

Plumbing Appliance Definition

CHANGE TYPE: Clarification

CHANGE SUMMARY: The definition of "plumbing appliance" has been changed to clarify the difference between appliances and fixtures.

2012 CODE: Plumbing Appliance. ~~Any one of a special class of plumbing fixtures~~ Water-connected or drain-connected devices intended to perform a special function. ~~Included are fixtures having the~~ These devices have their operation or control dependent on one or more energized components, such as motors, controls, or heating elements ~~or pressure or temperature sensing elements~~. Such ~~fixtures~~ devices are manually adjusted or controlled by the owner or operator, or are operated automatically through one or more of the following actions; a time cycle, a temperature range, a pressure range, a measured volume or weight.

CHANGE SIGNIFICANCE: The modified definition of "plumbing appliance" provides a better distinction between appliances and fixtures. The revised text updates and simplifies the definition, now recognizing these two classes as different. Examples of plumbing appliances include dishwashers, clothes washers, garbage disposals, water softeners, water purifiers, and water heaters.

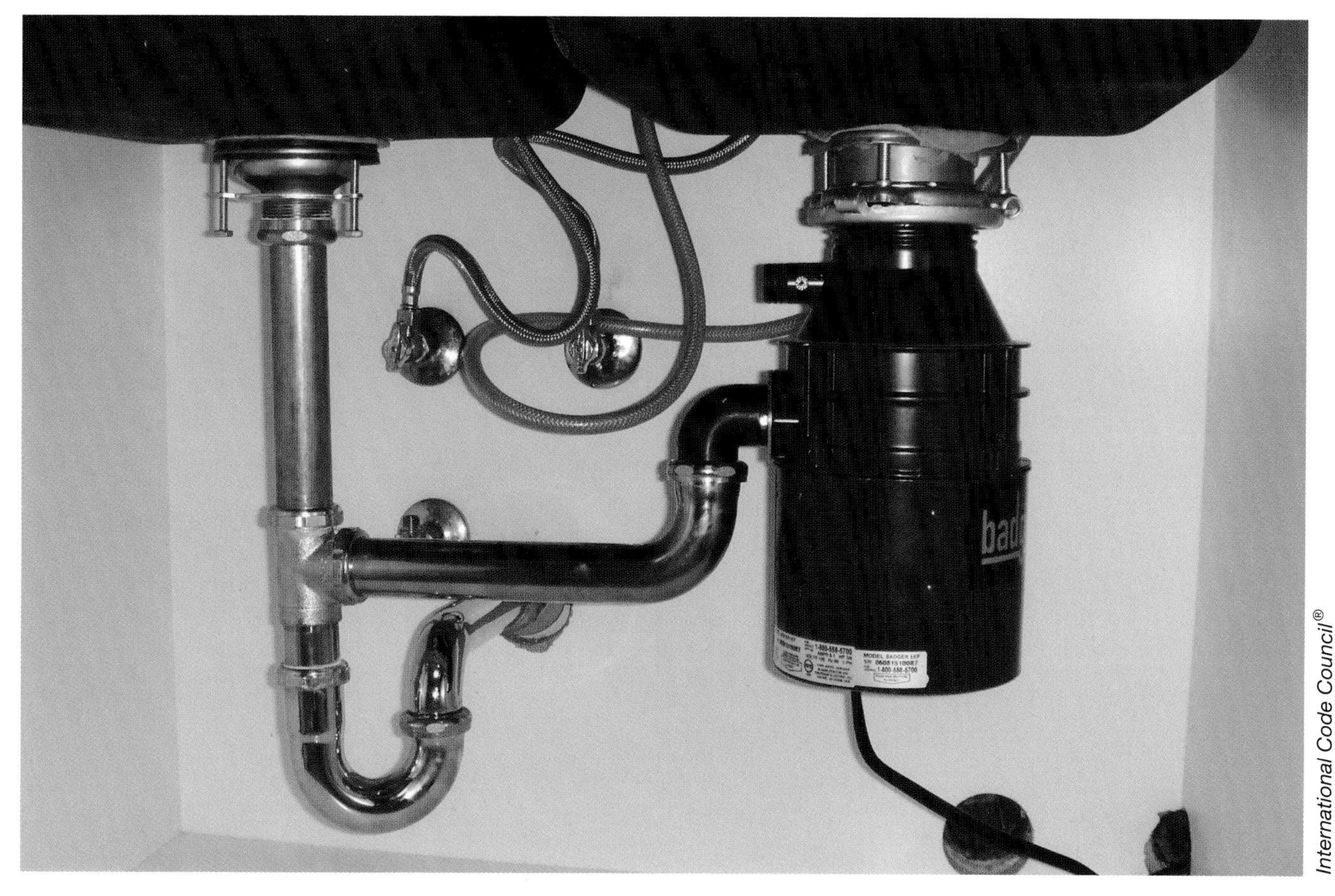

International Code Council®

Domestic garbage disposal

202

Grease Interceptor Definition

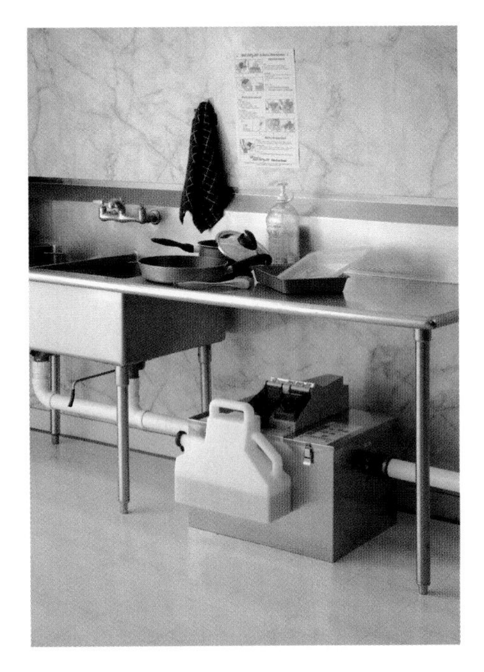

The installation of a typical hydromechanical grease interceptor *(Courtesy of Thermaco, Inc. 2010©)*

CHANGE TYPE: Modification

CHANGE SUMMARY: The definition of "grease interceptor" has been modified for consistency with current industry terms for the two general types of grease interceptors being used in plumbing installations.

2012 CODE: ~~**Grease Interceptor.** A plumbing appurtenance that is installed in a sanitary drainage system to intercept oily and greasy wastes from a wastewater discharge. Such device has the ability to intercept free-floating fats and oils.~~

Grease Interceptor.

Hydromechanical. Plumbing appurtenances that are installed in the sanitary drainage system to intercept free-floating fats, oils and grease from wastewater discharge. Continuous separation is accomplished by air entrainment, buoyancy and interior baffling.

Gravity. Plumbing appurtenances of not less than 500 gallons (1893 L) capacity that are installed in the sanitary drainage system to intercept free-floating fats, oils and grease from wastewater discharge. Separation is accomplished by gravity during a retention time of not less than 30 minutes.

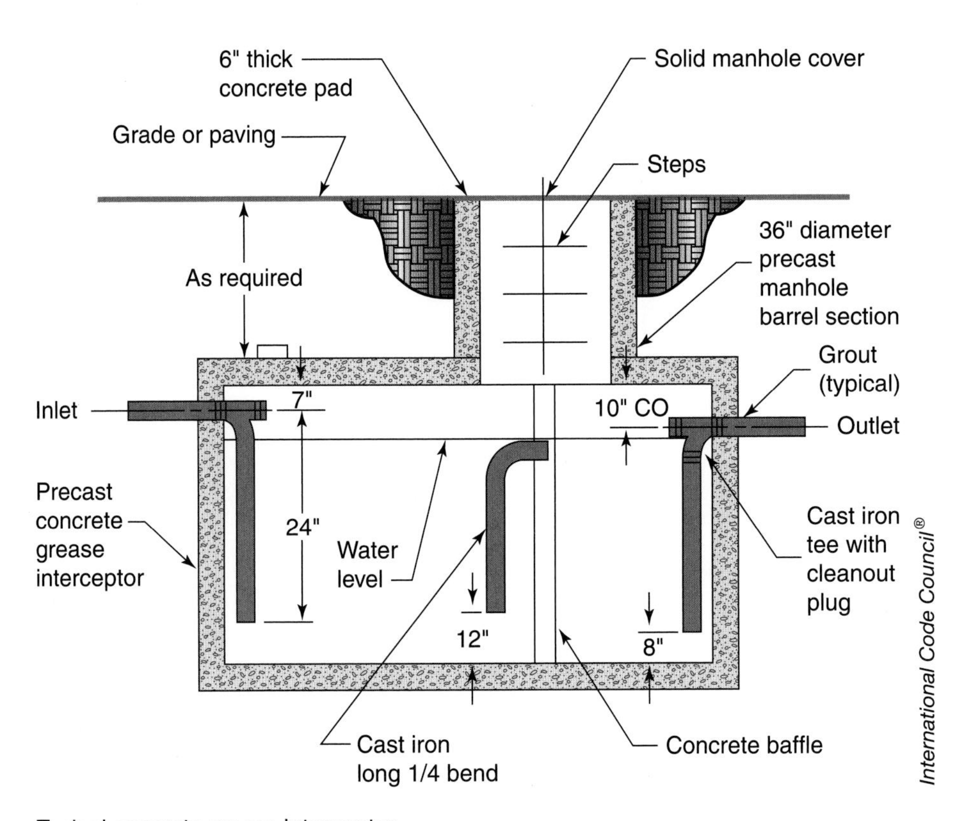

Typical concrete grease interceptor

CHANGE SIGNIFICANCE: The provisions of Section 1003.3.4 addressing grease interceptors and automatic grease removal devices were never intended to apply to gravity grease interceptors. The new terminology now makes a clear distinction between the two types of grease interceptors, hydromechanical and gravity. The revision to the definition along with the changes made in Section 1003.3.4 place the IPC in better alignment with product standards and industry terminology. Both types of grease interceptors require diligent effort by restaurant facility managers and staff to ensure that they are regularly maintained and properly serviced. For a grease collection device to work correctly, it must be properly designed, installed, maintained, and serviced regularly.

303.1, 303.4

Material Identification and Third-Party Certification

CHANGE TYPE: Clarification

CHANGE SUMMARY: The identification requirements for plumbing products and materials have been clarified.

2012 CODE: 303.1 Identification. Each length of pipe and each pipe fitting, trap, fixture, material, and device utilized in a plumbing system shall bear the identification of the manufacturer and any markings required by the applicable referenced standards.

303.4 Third-Party ~~Testing and~~ Certification. All plumbing products and materials shall ~~comply~~ be listed by a third-party certification agency as complying with the referenced standards~~, specifications and performance criteria of this code. and shall be identified in accordance with Section 303.1. When required by Table 303.4, plumbing products and materials shall either be tested by an approved third–party testing agency or certified by an approved third–party certification agency~~. Products and materials shall be identified in accordance with Section 303.1.

~~**TABLE 303.4 Products and Materials Requiring Third-Party Testing and Third-Party Certification**~~

~~Product or Material~~	~~Third-Party Certified~~	~~Third-Party Tested~~
~~Potable water supply system components and potable water fixture fittings~~	~~Required~~	–
~~Sanitary drainage and vent system components~~	~~Plastic pipe, fittings and pipe-related components~~	~~All Others~~
~~Waste fixture fittings~~	~~Plastic pipe, fittings and pipe-related components~~	~~All Others~~
~~Storm drainage system components~~	~~Plastic pipe, fittings and pipe-related components~~	~~All Others~~
~~Plumbing fixtures~~	–	~~Required~~
~~Plumbing appliances~~	~~Required~~	–
~~Backflow prevention devices~~	~~Required~~	–
~~Water distribution system safety devices~~	~~Required~~	–
~~Special waste system components~~	–	~~Required~~
~~Subsoil drainage system components~~	–	~~Required~~

PEX tubing label with complete identification and conformance information *(Courtesy of Uponor Corporation)*

CHANGE SIGNIFICANCE: The modification to the identification provisions of Section 303.1 clarifies the intent of the code that products and materials shall bear the identification of the manufacturer, as well as the identification requirements that are referenced by the applicable standard. As a result of the modifications made to Section 303.4, all plumbing products and materials must now be listed by a third-party certification agency. Table 303.4 was deleted as a result of the modifications made to Section 303.4.

308.9

Parallel Water Distribution Systems

CHANGE TYPE: Modification

CHANGE SUMMARY: In parallel water distribution systems, the hot and cold water piping may now be grouped in the same pipe bundle.

2012 CODE: **308.9 Parallel Water Distribution Systems.** Piping bundles for manifold systems shall be supported in accordance with Table 308.5. Support at changes in direction shall be in accordance with the manufacturer's ~~installation~~ instructions. Where ~~H~~hot ~~and cold~~ water piping ~~shall not be grouped in the same~~ is bundled with cold or hot water piping, each hot water pipe shall be insulated.

CHANGE SIGNIFICANCE: Hot water piping is now permitted to be bundled together with other cold or hot water piping. However, this installation method is only permitted where each hot water pipe is insulated. Only one of the hot water pipes in a parallel water distribution system actually has hot water running through it at any given time. The other hot water pipes have cold water sitting in them and heat transfer between the hot pipe being used and the other piping will take place. Manifold plumbing systems are control centers for hot and cold water that feed flexible PEX supply lines to individual fixtures. Cold water and un-insulated hot water piping in the same bundles are known to absorb large amounts of heat. This code modification will prevent this costly heat transfer.

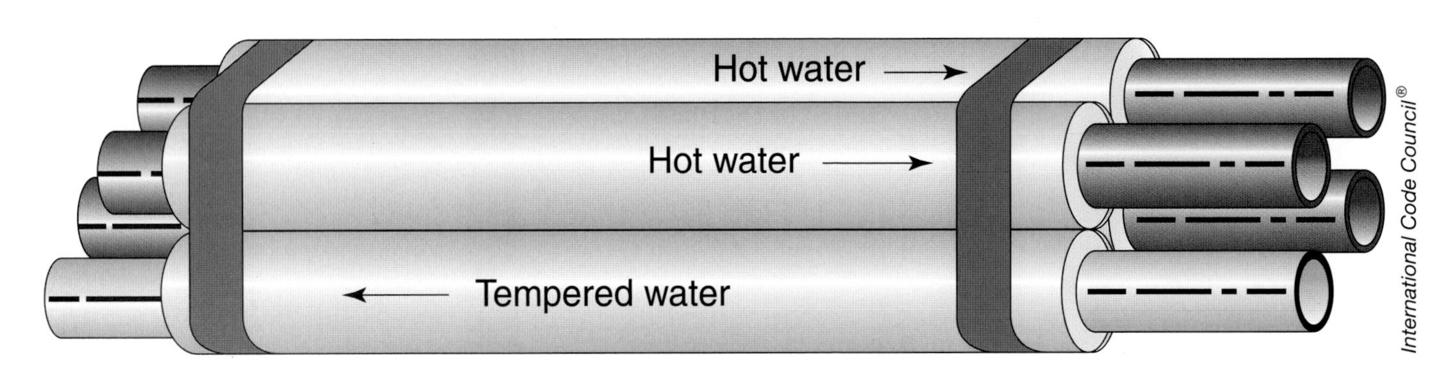

Example of insulated bundle piping

CHANGE TYPE: Modification

315.1

Sealing of Annular Spaces at Penetrations

CHANGE SUMMARY: The provisions for sealing any annular spaces created at piping penetrations have been revised to be consistent with the building envelope sealing requirements of the *International Energy Conservation Code.*

2012 CODE: **~~305.4~~ 315.1 ~~Sleeves~~ Sealing of Annular Spaces.** The annular space~~s~~ between the outside of a pipe and the inside of a pipe sleeve~~s~~, ~~and pipes~~ or between the outside of a pipe and an opening in a building envelope wall, floor, or ceiling assembly penetrated by a pipe shall be ~~filled or tightly caulked~~ sealed in an approved manner with caulking material or closed with a gasketing system. The caulking material, foam sealant, or gasketing system shall be designed for the conditions at the penetration location and shall be compatible with the pipe, sleeve and building materials in contact with the sealing materials. Annular spaces ~~between~~ created by pipes penetrating ~~sleeves and pipes in~~ fire resistance-rated assemblies or membranes of such assemblies shall be ~~filled or tightly caulked~~ sealed or closed in accordance with Section 714 of the *International Building Code.*

315.1 continues

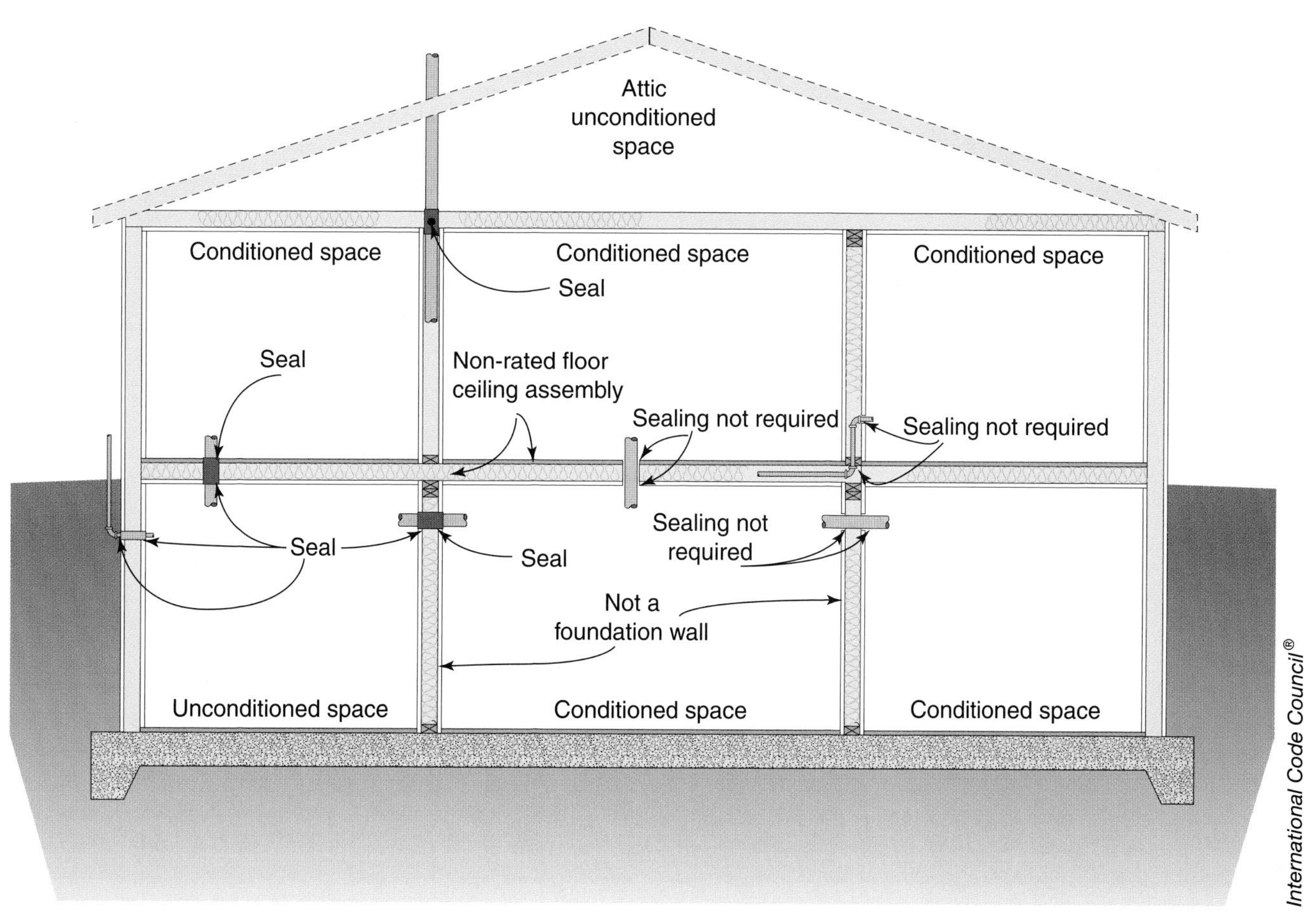

Sealing of building envelope

315.1 continued

CHANGE SIGNIFICANCE: It is important that penetrations of the building envelope and other building elements be appropriately sealed. The modified text clarifies that only the ends of the annular spaces need to be sealed or closed. Filling of the entire annular space cavity is unnecessary for preventing uncontrolled air movement. The term "tightly caulked" was removed, as it was considered outdated language from the era of "packing and pouring" lead joints. It is anticipated that there will be only a few situations that would warrant having a pipe so rigidly fixed in a through-penetration. The term "pipe" was added in the reference to sleeves in order to clarify which sleeves are to be considered.

Clarification has been given to what was sometimes interpreted to require sealing between pipe and flexible plastic sleeving that is used for corrosion protection. It was also considered important to add the requirements that sealing materials be compatible with all items that they might come in contact with and that the materials be suitable for the weather and temperature conditions of the application. There are several solvent-based caulking materials that affect plastic piping, and there are instances where a caulking material is inappropriate for outdoor conditions, resulting in rainwater damage to the building.

An additional change made was to clarify and emphasize the importance of ensuring that where fire-resistance-rated assemblies are being penetrated by pipes, specific materials and methods in accordance with the *International Building Code* (IBC) must be used. Proper firestopping methods are critical for fire safety.

Table 403.1

Minimum Number of Required Plumbing Fixtures

CHANGE TYPE: Modification

CHANGE SUMMARY: Service sinks are no longer required in Group B and M occupancies where the occupant load does not exceed 15.

2012 CODE:

TABLE 403.1 (IBC TABLE 2902.1) Minimum Number of Required Plumbing Fixtures[a] (See Sections 403.2 and 403.3)

No.	Classification	Occupancy	Description	Other
2	Business	B	Buildings for the transaction of business, professional services, other services involving merchandise, office buildings, banks, light industrial and similar uses	1 service sink[g]
6	Mercantile	M	Retail stores, service stations, shops, salesrooms, markets and shopping centers	1 service sink[g]

g. For business and mercantile occupancies with an occupant load of 15 or fewer, service sinks shall not be required.

***Reader's Note:** *Other changes may occur in Table 403.1 that will be addressed in different areas of this book; those portions of the table not addressed remain unchanged.*

CHANGE SIGNIFICANCE: A new allowance limited to business and mercantile occupancies permits the omission of a service sink where the occupant load of the establishment is 15 or less persons. The basis for the exception is that there are other thresholds established within the code that provide for reduced requirements where the maximum occupancy is very low. For example, the requirement for separate male and female restrooms (separate facilities for each sex) is only applicable where there are more than 15 occupants, increasing to 100 occupants in mercantile sales occupancies. The allowance granted by footnote "g" eliminates the mandate for a service sink in small business and mercantile occupancies. In a small facility, such as a retail store with a sales area of not more than 3,000 square feet or an office with a maximum floor area of 1,500 square feet, a service sink and the associated closet can occupy a disproportionate amount of floor space. Typically, service sinks in these small occupancies are rarely, if ever, used.

403.2

Separate Toilet Facilities in Group M Occupancies

CHANGE TYPE: Modification

CHANGE SUMMARY: The exemption from separate plumbing facilities for each sex in Group M mercantile occupancies now applies where the occupant load of the occupancy does not exceed 100.

2012 CODE: **403.2 Separate Facilities.** Where plumbing fixtures are required, separate facilities shall be provided for each sex.

Exceptions:

1. Separate facilities shall not be required for dwelling units and sleeping units.
2. Separate facilities shall not be required in structures or tenant spaces with a total occupant load, including both employees and customers, of 15 or less.
3. Separate facilities shall not be required in mercantile occupancies in which the maximum occupant load is ~~50~~ 100 or less.

CHANGE SIGNIFICANCE: Recent years have seen an increase in mixed-use buildings that are predominantly residential in use with one or more small, secondary retail components. Such retail spaces are quite often developed into neighborhood retail that is boutique in nature and classified as a Group M occupancy. A study by the U.S. Department of Labor, Bureau of Labor Statistics indicated that the typical floor area of the retail units fell within a range of 1,500 square feet to 3,000 square feet. As an example, based on *International Building Code* (IBC) Table 1004.1.1, at 30 square feet per occupant, a typical space of 2,400 square feet would

International Code Council®

Group M occupancy

have an occupant load of 80 persons. Consequently, such a space did not previously qualify for Exception 3 of the IPC Section 403.5 that allows for a single toilet facility to serve up to 50 occupants. Therefore, the space would have required separate toilet facilities for males and females. The requirement for separate facilities for each sex placed on these smaller retail occupancies often led to manipulation of the occupant load calculation for the purpose of avoiding the additional toilet facility. Because a Group M occupancy requires a second exit where the occupant load exceeds 49, the manipulation of occupant load subsequently adversely impacted the means-of-egress requirements for the space. By increasing the occupant load threshold to 100 persons, separate toilet facilities for each sex are not required for those small retail spaces having floor areas not greater than 3,000 square feet.

Two accessible single-user toilet facilities occupy approximately 80 square feet of floor area. In a 1,500-square-foot tenant space, these facilities would occupy more than 5% of the total space. The increase in the occupant load threshold now allows for the industry norm in boutique retail tenant size to be accommodated with one single-user toilet facility. Given that it is rare that these small retail spaces would be occupied by the number of persons equal to the design occupant load, and that IPC Table 403.1 indicates that two water closets are permitted to serve up to 1,000 persons, the provision of a single toilet facility appears to be more than adequate for the size of space that the proposed occupant load threshold increase would allow. The limited floor area of 3,000 square feet that the proposed occupant load threshold can accommodate is such that neither a 500- nor 300-foot travel distance limitation as required in IPC Section 403.3 would ever be exceeded.

403.2.1

Family or Assisted-Use Toilet Facilities Serving as Separate Facilities

International Code Council®

Family or assisted-use toilet facility

CHANGE TYPE: Addition

CHANGE SUMMARY: Where separate toilet facilities for each sex are required and only one water closet is mandated in each facility, two family or assisted-use toilet facilities are now permitted to substitute for the separate facilities for each sex.

2012 CODE: **403.2.1 Family or Assisted-Use Toilet Facilities Serving as Separate Facilities.** Where a building or tenant space requires a separate toilet facility for each sex and each toilet facility is required to have only one water closet, two family/assisted-use toilet facilities shall be permitted to serve as the required separate facilities. Family or assisted-use toilet facilities shall not be required to be identified for exclusive use by either sex as required by Section 403.4.

CHANGE SIGNIFICANCE: Separate toilet facilities for males and females are required in most buildings. In many cases, only one water closet is mandated within each toilet facility. In such situations, the code now permits the substitution of two family/assisted-use toilet rooms in lieu of the two separate toilet rooms for each sex. The advantage of allowing two family/assisted-use toilet facilities to serve as the required separate facilities is the efficiency provided when either toilet room can be used by either sex. This increases the availability of facilities in smaller spaces without needing to offer multi-user toilet facilities. There are often situations where a single gender-based toilet facility can be unavailable for periods of up to 15 minutes when, for example, the current occupant is using it for companion care, to change diapers, or to change a colostomy bag. There will also be less of an impact to potential users when one toilet room is being cleaned or serviced. This is not a new concept, as the IPC has always permitted the use of a shared toilet room in buildings or tenant spaces with low occupant loads. As another example, males and females use the same toilet facility on airplanes.

403.3.2

Relationship of Toilet Rooms and Food Preparation Areas

CHANGE TYPE: Addition

CHANGE SUMMARY: The IBC requirement prohibiting the opening of toilet rooms directly into food preparation areas is now also established in the IPC.

2012 CODE: **403.3.1 Toilet Room Ingress and Egress.** Toilet rooms shall not open directly into a room used for the preparation of food for service to the public.

CHANGE SIGNIFICANCE: IBC Section 1210.5 has historically prohibited openings between a toilet room and any room or space where food is being prepared for the public, such as a commercial kitchen that serves a restaurant dining area. The requirement that toilet rooms not open directly into rooms where food is prepared for the public is necessary to keep the food preparation areas in a sanitary condition. Replicating the building code provision in the IPC will be helpful and increase efficiency for plumbing designers, installers, inspectors, and other IPC users.

403.3.2 continues

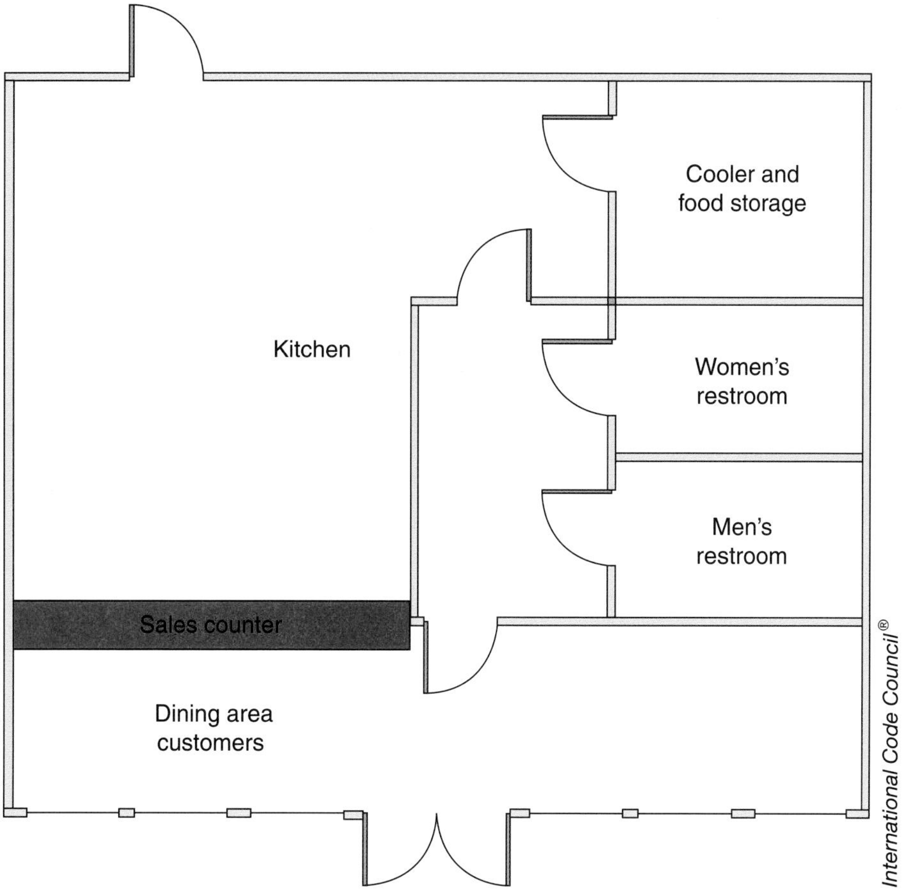

Cooler and food storage

Kitchen

Women's restroom

Men's restroom

Sales counter

Dining area customers

International Code Council®

Acceptable condition

403.3.2 continued

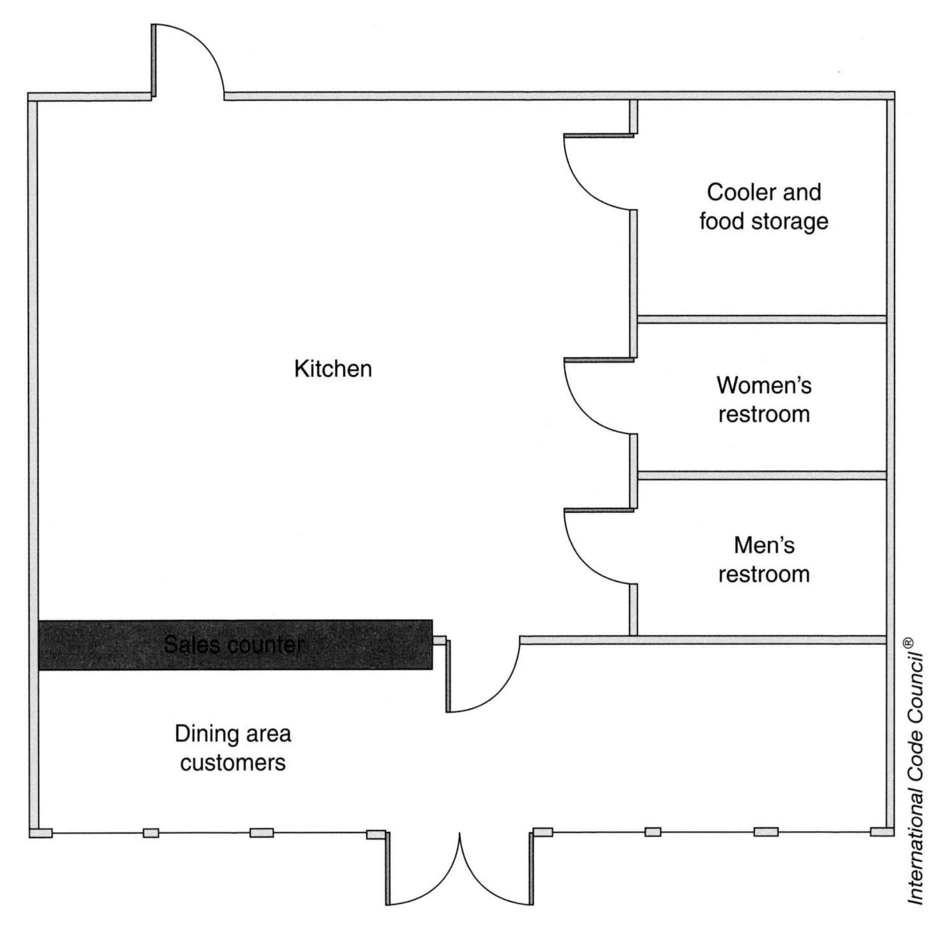

Unacceptable condition

403.3.6

Locking of Toilet Room Doors

CHANGE TYPE: Addition

CHANGE SUMMARY: Locking devices are now specifically prohibited on the egress door of toilet rooms designed for multiple occupants.

2012 CODE: **403.3.6 Door Locking.** Where a toilet room is designed for multiple occupants, the egress door for the room shall not be lockable from the inside of the room. This section does not apply to family or assisted-use toilet rooms.

CHANGE SIGNIFICANCE: The doors of multiple-occupant toilet rooms must no longer be capable of being locked from the inside of the room. Restricting the egress door in this way will reduce the possibility of inappropriate activities that are more likely to occur when an occupant can restrict entry to the toilet room. Such locking potential can also restrict immediate egress from the toilet room when it may be necessary.

International Code Council®

Acceptable nonlockable door installation for a multiple-occupant toilet room

403.5

Drinking Fountain Locations

CHANGE TYPE: Addition

CHANGE SUMMARY: Where drinking fountains are required, the permitted locations of the fountains have been specified regarding their placement in multi-tenant facilities, similar to the permitted locations for required public toilet facilities.

2012 CODE: **403.5 Required Drinking Fountains.** Drinking fountains shall not be required to be located in individual tenant spaces provided that public drinking fountains are located within a travel distance of 500 feet of the most remote location in the tenant space and not more than one story above or below the tenant space. Where the tenant space is in a covered or open mall, such distance shall not exceed 300 feet. Drinking fountains shall be located on an accessible route.

CHANGE SIGNIFICANCE: The sharing of public restroom facilities in multi-tenant facilities has historically been permitted under Section 403.3 of the IPC, but the code was silent on the sharing of drinking fountains. The new provision recognizes that if employees and the public can share public restroom facilities, then they should be able to also share drinking fountains if located within a reasonable distance. The travel distance restriction of 500 feet maximum between a public drinking fountain and the most remote location in the tenant space, as well as the limitation requiring placement of the fountain not more than one story above or below the tenant space, is almost identical to the language used in Section 403.3.2 for toilet facilities. The limiting distance of 300 feet in covered mall buildings is the same distance required for toilet facilities.

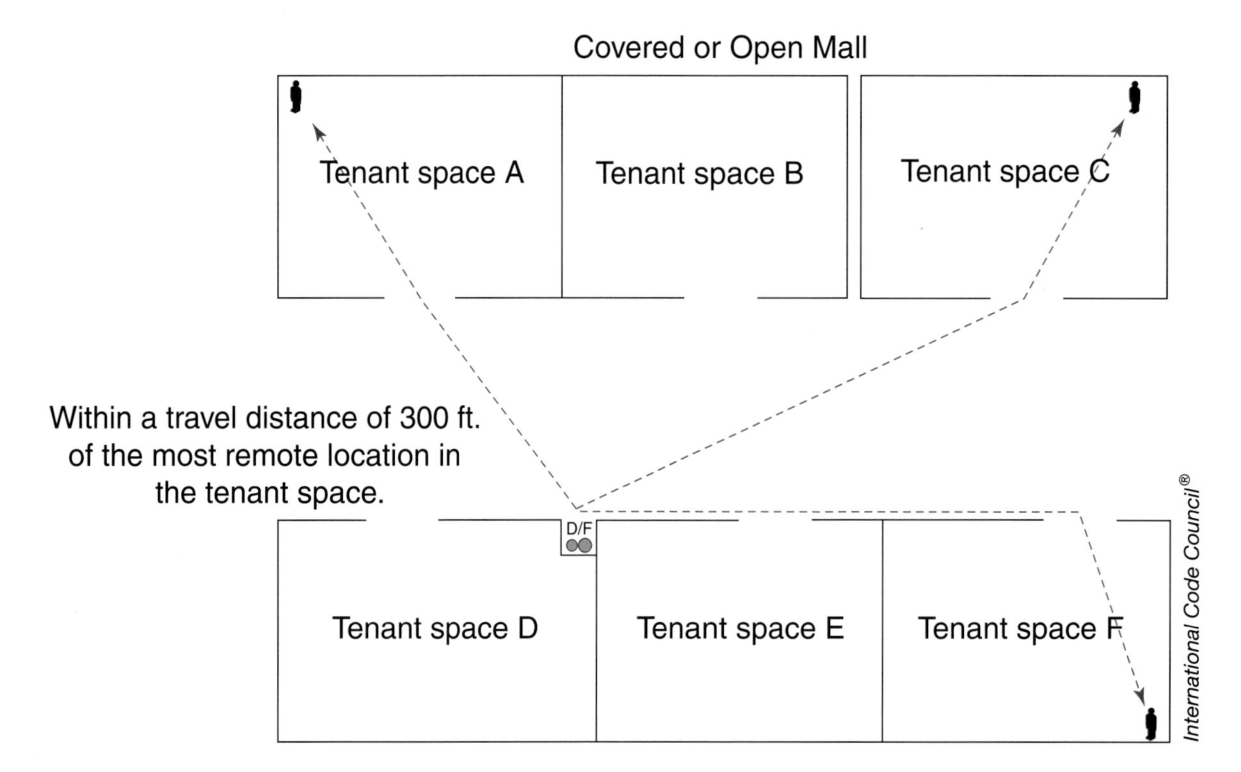

Drinking fountain location

405.3.1

Minimum Water Closet Compartment Size

CHANGE TYPE: Modification

CHANGE SUMMARY: The minimum depth of a water closet compartment containing a wall-hung water closet has been reduced from 60 inches to 56 inches.

2012 CODE: 405.3.1 Water Closets, Urinals, Lavatories, and Bidets. A water closet, urinal, lavatory or bidet shall not be set closer than 15 inches (381 mm) from its center to any side wall, partition, vanity or other obstruction, or closer than 30 inches (762 mm) center to center between adjacent fixtures. There shall be at least a 21-inch (533-mm) clearance in front of the water closet, urinal, lavatory or bidet to any wall, fixture or door. Water closet compartments shall be not less than 30 inches (762 mm) in width and 60 inches (1524 mm) in depth for floor mounted water closets and not less than 30 inches (762 mm) in width and 56 inches (1422 mm) in depth for wall hung water closets ~~(see Figure 405.3.1)~~.

CHANGE SIGNIFICANCE: A wall-hung nonaccessible water closet compartment is now permitted to be 56 inches in depth, which is 4 inches shorter in length than required for a compartment containing a floor-mounted water closet. The IPC is now closer in alignment with the accessible water closet compartment depth requirements set forth in Section 604.8.2 of ICC/ANSI A117.1 2009, where wall-hung water closets are provided in accessible water closet compartments.

405.3.1 continues

405.3.1 continued

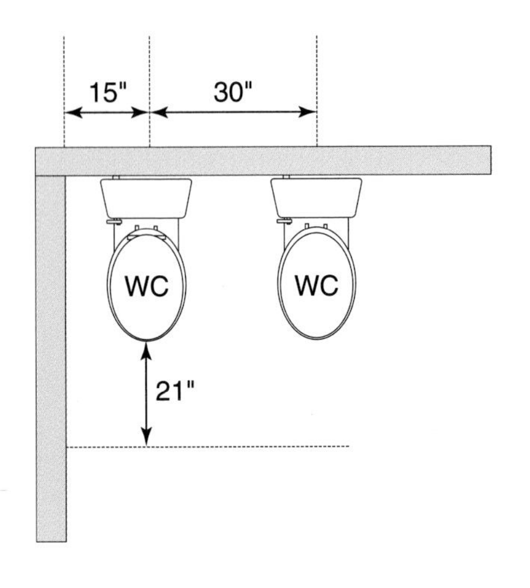

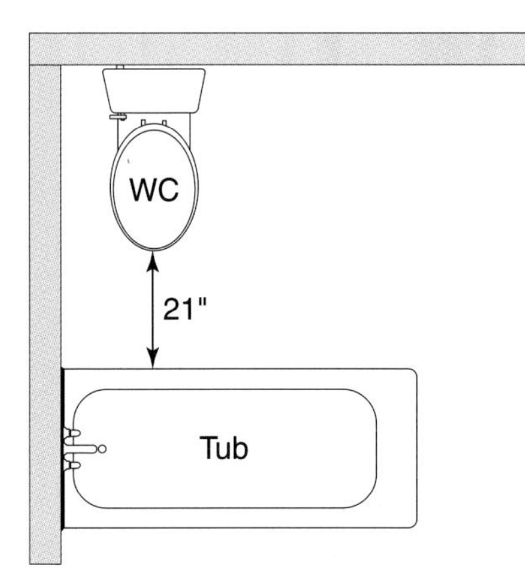

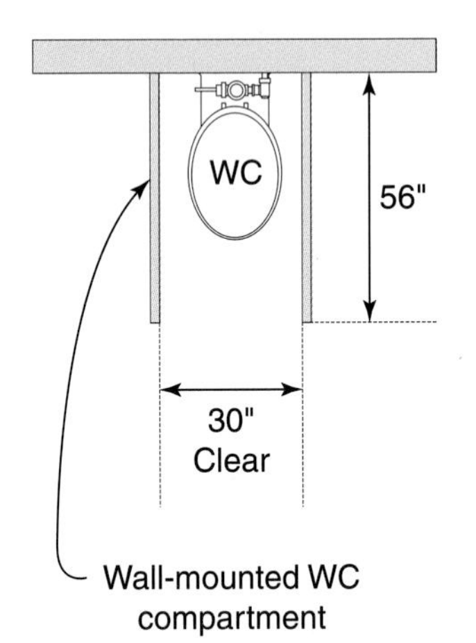

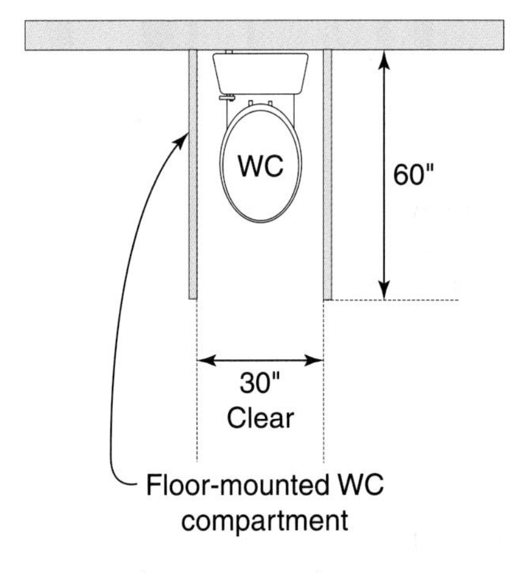

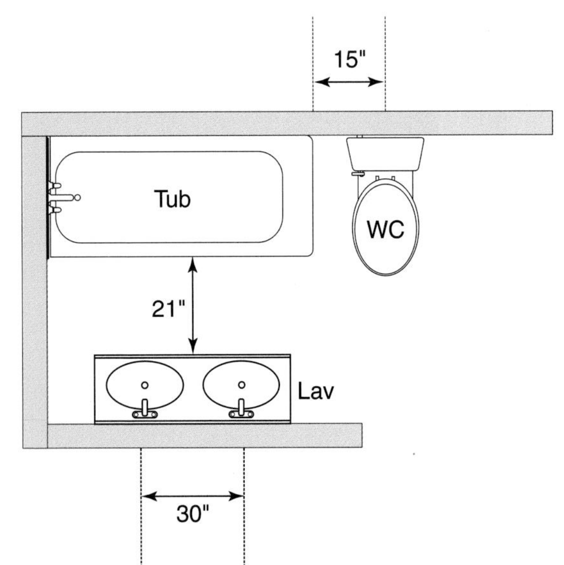

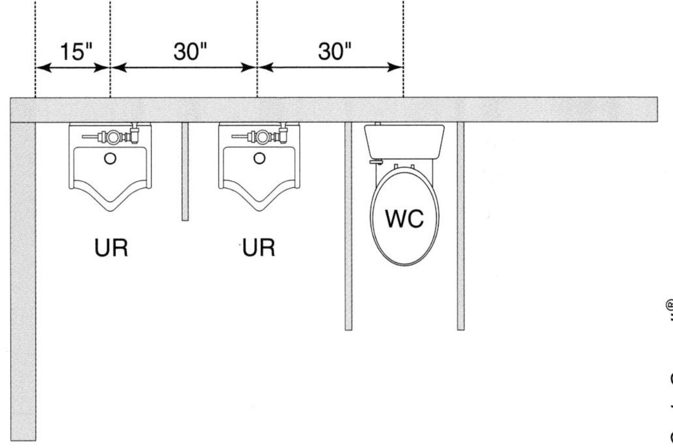

Fixture clearance

405.4

Floor and Wall Drainage Connections

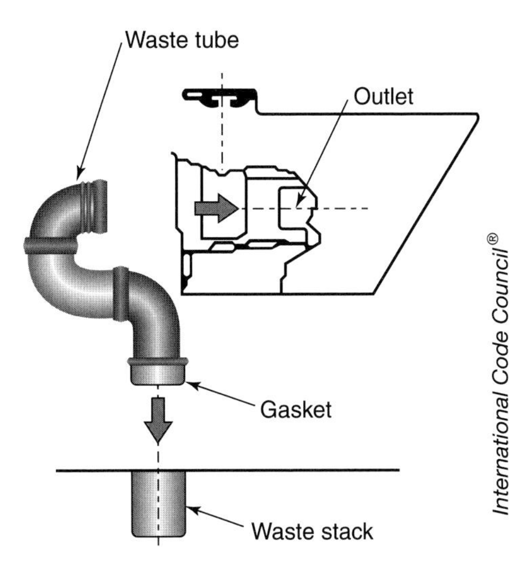

Gasketed waste tube connection for water closet

CHANGE TYPE: Modification

CHANGE SUMMARY: The use of a waste connector and sealing gasket is now permitted as an acceptable means to connect floor outlet plumbing fixtures, allowing for water closet installations that are provided with a gasketed waste tube outlet connection.

2012 CODE: 405.4 Floor and Wall Drainage Connections. Connections between the drain and floor outlet plumbing fixtures shall be made with a floor flange or a waste connector and sealing gasket. The waste connector and sealing gasket joint shall comply with the joint tightness test of ASME A112.4.3 and shall be installed in accordance with the manufacturer's installation instructions. The flange shall be attached to the drain and anchored to the structure. Connections between the drain and wall-hung water closets shall be made with an *approved* extension nipple or horn adaptor. The water closet shall be bolted to the hanger with corrosion-resistant bolts or screws. Joints shall be sealed with an *approved* elastomeric gasket, flange-to-fixture connection complying with ASME A112.4.3 or an *approved* setting compound.

CHANGE SIGNIFICANCE: Historically, a "flanged" outlet connection for floor-mounted water closets has been the only acceptable method for making the connection between the drain and a floor outlet plumbing fixture. Connections of this type are typical of water closets designed for the North American market. The recognition of a waste connector and sealing gasket allows for another acceptable type of water closet connection method that will make more water closet products available to designers and installers. The new allowance recognizes this commonly used international method of connection. The connection arrangement consists of a waste tube connector on the water closet that is inserted into an elastomeric gasket. The waste tube and gasket are then inserted into the drain pipe opening at the floor line, and the gasket provides the seal between the water closet's waste tube and the drain pipe. The water closet fixture is then anchored directly to the floor using mounting brackets or fasteners. These anchors are often concealed to allow for a smooth, sanitary exterior interface to the floor.

This design is used almost exclusively in Europe and other locations worldwide, and offers many advantages over wax-ring flange seals. ASME A112.4.3, a standard already referenced in the IPC and the International Residential Code (IRC), requires that the connection be leak-tight to pressures up to 10 psi. Such water closet designs are available in a wide range of rough-in dimensions.

407.2

Bathtub Waste Outlets and Overflows

CHANGE TYPE: Modification

CHANGE SUMMARY: Bathtubs are now required to be equipped with an overflow, and the required stopper must be watertight.

2012 CODE: **407.2 Bathtub Waste Outlets and Overflows.** Bathtubs shall ~~have~~ be equipped with a waste outlet~~s~~ and an overflow outlet. The ~~minimum~~ outlets shall ~~of~~ be connected to waste tubing or piping not less than 1 ½ inches (38 mm) in diameter. ~~, and~~ ~~t~~The waste outlet shall be equipped with a~~n approved~~ watertight stopper.

CHANGE SIGNIFICANCE: Even though most bathtubs are installed with overflows, the code text has not previously been clear as to whether or not an overflow was required. New language specifically states that bathtubs must be provided with an overflow outlet. Overflows for bathtubs are a safeguard to prevent flooding and are now required to be installed in addition to the waste outlet. The insertion of the term "watertight" regarding the required waste outlet stopper clarifies that the purpose of the stopper is to allow the tub to be filled with water and hold the water in place.

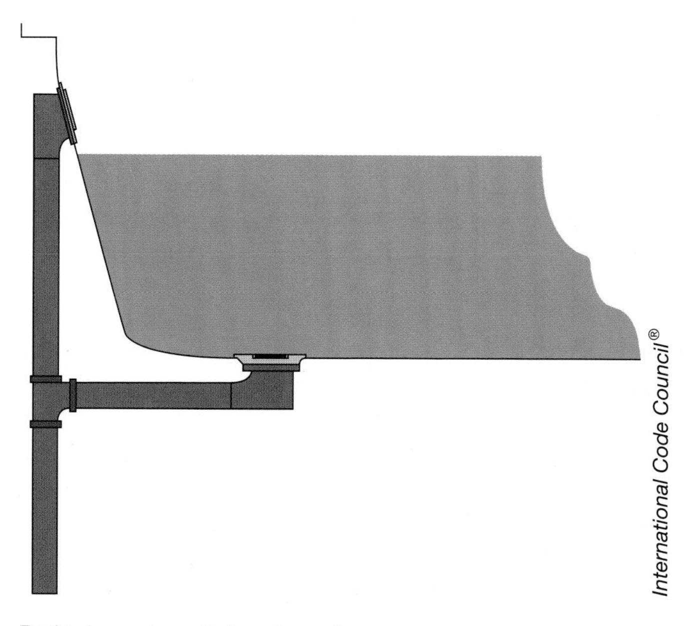

Bathtub waste outlet and overflow

410

Minimum Required Number of Drinking Fountains

CHANGE TYPE: Modification

CHANGE SUMMARY: The IBC provisions addressing the minimum required number of drinking fountains have been replicated in the IPC to provide clarity and consistency of application.

2012 CODE: 410.1 Approval. Drinking fountains shall conform to ASME A112.19.1~~M~~/CSA B45.2 or ASME A112.19.2~~M~~/CSA B45.1 ~~or ASME A112.19.9M~~ and water coolers shall conform to ARI 1010. Drinking fountains and water coolers shall conform to NSF 61, Section 9. ~~Where water is served in restaurants, drinking fountains shall not be required. In other occupancies, where drinking fountains are required, water coolers or bottled water dispensers shall be permitted to be substituted for not more than 50 percent of the required drinking fountains.~~

410.2 Minimum Number. Not fewer than two drinking fountains shall be provided. One drinking fountain shall comply with the requirements for people who use a wheelchair and one drinking fountain shall comply with the requirements for standing persons.

> **Exception:** A single drinking fountain that complies with the requirements for people who use a wheelchair and standing persons shall be permitted to be substituted for two separate drinking fountains.

410.3 Substitution. Where restaurants provide drinking water in a container free of charge, drinking fountains shall not be required in those restaurants. In other occupancies, where drinking fountains are required, water coolers or bottled water dispensers shall be permitted to be substituted for not more than 50 percent of the required number of drinking fountains.

CHANGE SIGNIFICANCE: Although the minimum required number of drinking fountains for each type of occupancy has historically been addressed in the IPC, Section 1109.5 of the IBC went into more detail concerning the requirements for people who use a wheelchair. The IBC criteria have now been brought into the IPC to give it more clarity and to provide consistency with the IBC. No fewer than two drinking fountains shall be provided where a drinking fountain is required. At least one drinking fountain shall comply with the requirements for people who use a wheelchair and at least one drinking fountain shall comply with the requirements for standing persons. The seated and standing drinking fountains that serve a facility need not be provided at the same location in the facility. The exception allows for the use of a single fixture that accommodates both seated and standing persons. Technical criteria for both wheelchair-accessible fountains and standing-person fountains are located in Section 602 of ICC A117.1.

Restaurants have historically been exempted from the requirement for drinking fountains, provided water service is available to the customers. This exemption is now only applicable where the water is provided in a container free of charge. Those restaurants that only provide water at a cost to the customer must have complying drinking fountains.

410 continues

410 continued

Where the plumbing code requires this number of drinking fountains:

The building code requires either of these configurations:

Combo "Hi-Lo Units"

or

Combo "Hi-Lo Units"

or

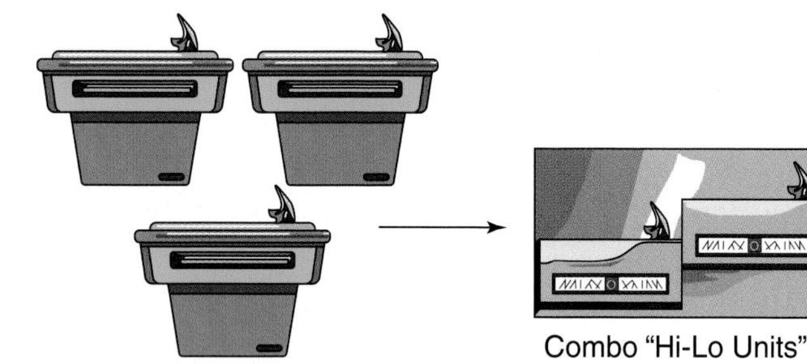

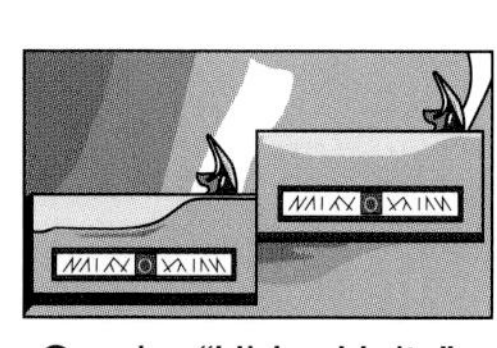

Combo "Hi-Lo Units"

+

or

Minimum required number of drinking fountains

International Code Council®

417.5.2.6

Shower Pan Liner Materials

CHANGE TYPE: Addition

CHANGE SUMMARY: Recognition of an acceptable shower pan liner system using liquid-type, trowel-applied, load-bearing, bonded waterproof materials has been added to the current listing of acceptable shower floor liner methods.

2012 CODE: **417.5.2.6 Liquid Type, Trowel Applied, Load Bearing, Bonded Waterproof Materials.** Liquid type, trowel applied load bearing, bonded waterproof materials shall meet the requirements of ANSI A118.10 and shall be applied in accordance with the manufacturer's installation instructions.

CHANGE SIGNIFICANCE: Another acceptable form of waterproofing has been recognized for the on-site construction of shower floors. The membrane that is applied is a thin, load-bearing waterproofing designed specifically for the special requirements of ceramic tile, stone, and brick installations. A self-curing liquid rubber polymer and a reinforcing fabric are quickly applied to form a flexible, seamless waterproofing membrane that bonds to a wide variety of substrates. The membrane must be applied in accordance with the manufacturer's installation instructions.

ANSI A118.10 waterproof membrane
(Courtesy of Laticrete International)

424.9

Water Closet Personal Hygiene Devices

CHANGE TYPE: Addition

CHANGE SUMMARY: The recognition of performance standard ASME A112.4.2 now ensures the protection of the public by setting temperature limits and minimum acceptable backflow protection requirements for water closet personal hygiene devices.

2012 CODE: **424.9 Water Closet Personal Hygiene Devices.** Personal hygiene devices integral to water closets or water closet seats shall conform to the requirements of ASME A112.4.2.

Chapter 14:

ASME

A112.4.2-2003 (R2008) Water Closet Personal Hygiene Devices

CHANGE SIGNIFICANCE: The American Society for Mechanical Engineers (ASME) consensus standard for water closet personal hygiene devices, A112.4.2, is now listed in the IPC in order to help regulate these devices. This standard establishes general and performance requirements, test methods, and marking requirements for bidet sprays and other optional features as applied to water closets, water closet seats, and other retrofit devices. The application of the provisions within the standard will ensure protection of plumbing systems from backflow and protect public safety by limiting the temperature of the water dispensed.

Toto Washlet C110 *(Courtesy of Toto USA, Inc.)*

504.4.1

Water Heater Storage Tank Relief Valves

CHANGE TYPE: Clarification

CHANGE SUMMARY: It has been clarified that water heaters with separate storage tanks shall be provided with complying temperature and pressure protection.

2012 CODE: 504.4.1 Installation. Such valves shall be installed in the shell of the water heater tank. Temperature relief valves shall be so located in the tank as to be actuated by the water in the top 6 inches (152 mm) of the tank served. For installations with separate storage tanks, the approved, self-closing (levered) pressure relief valve and temperature relief valve or combination thereof conforming to ANSI Z21.22 valves shall be installed on ~~the tank and there shall not be any type of valve installed between the water heater and the storage tank.~~ both the storage water heater and storage tank. There shall not be a check valve or shutoff valve between a relief valve and the heater or tank served.

CHANGE SIGNIFICANCE: Temperature and pressure protection has historically been addressed in detail for storage water heaters, while the protection requirements for water heaters with separate storage tanks were limited. Now both storage water heaters and those water heaters with separate storage tanks are to have the same safety level of protection and must conform to ANSI Z21.22. A valve is permitted between the water heater and separate storage tank to aid in servicing, maintenance, or replacement.

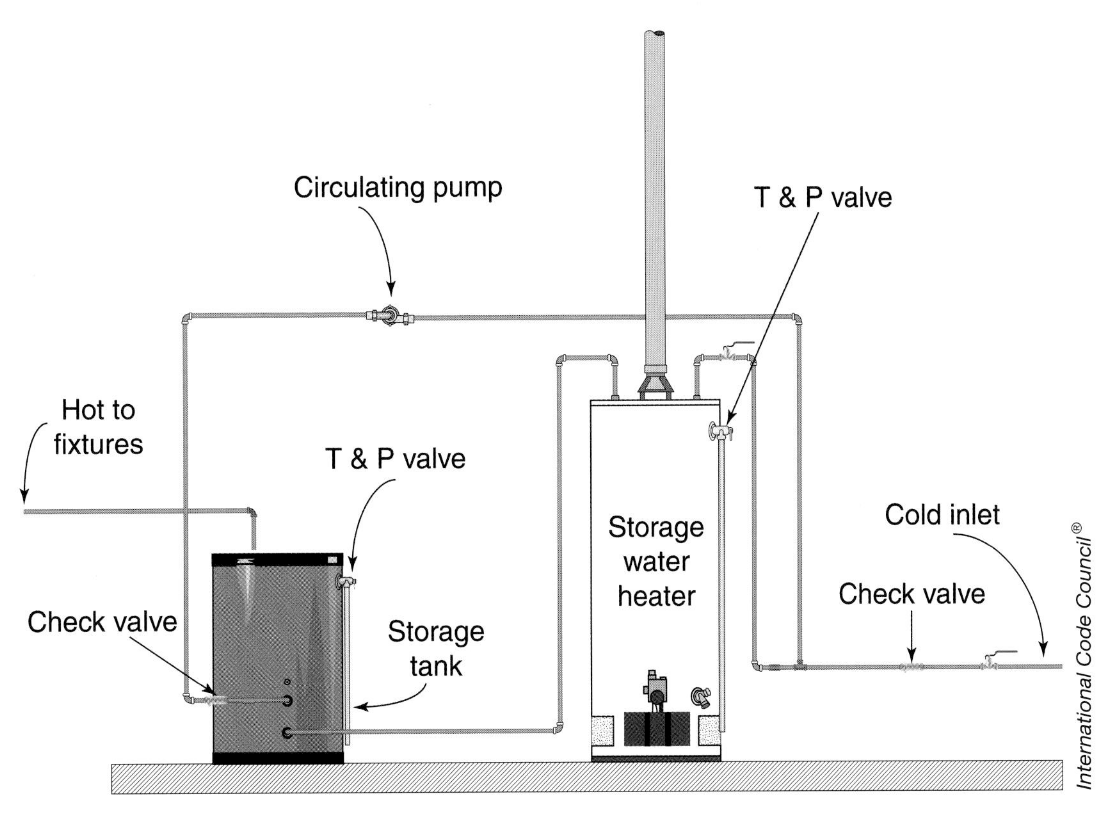

Water heater and storage tank installation

504.7

Water Heater Pans

CHANGE TYPE: Modification

CHANGE SUMMARY: It has been clarified that drain pans are only required for storage-tank-type water heaters or hot water storage tanks.

2012 CODE: 504.7 Required Pan. Where a storage tank-type water heater~~s~~ or a hot water storage tank~~s~~ ~~are~~ is installed in a location~~s~~ where water leakage from ~~of~~ the tank~~s~~ ~~or connections~~ will cause damage, the tank ~~or water heater~~ shall be installed in a galvanized steel pan having a material thickness of not less than 0.236 inch (0.6010 mm) (No. 24 gage), or other pans approved for such use.

CHANGE SIGNIFICANCE: Previously, it was unclear whether or not a tankless-type water heater needs to be provided with a pan. A tankless water heater does not have a storage tank and thus does not present any greater risk of water leakage than piping in a water distribution system that has been installed and pressure tested in accordance with the code. As a result, the provision was revised to make the pan requirement specific to only storage-tank water heaters and hot water storage tanks.

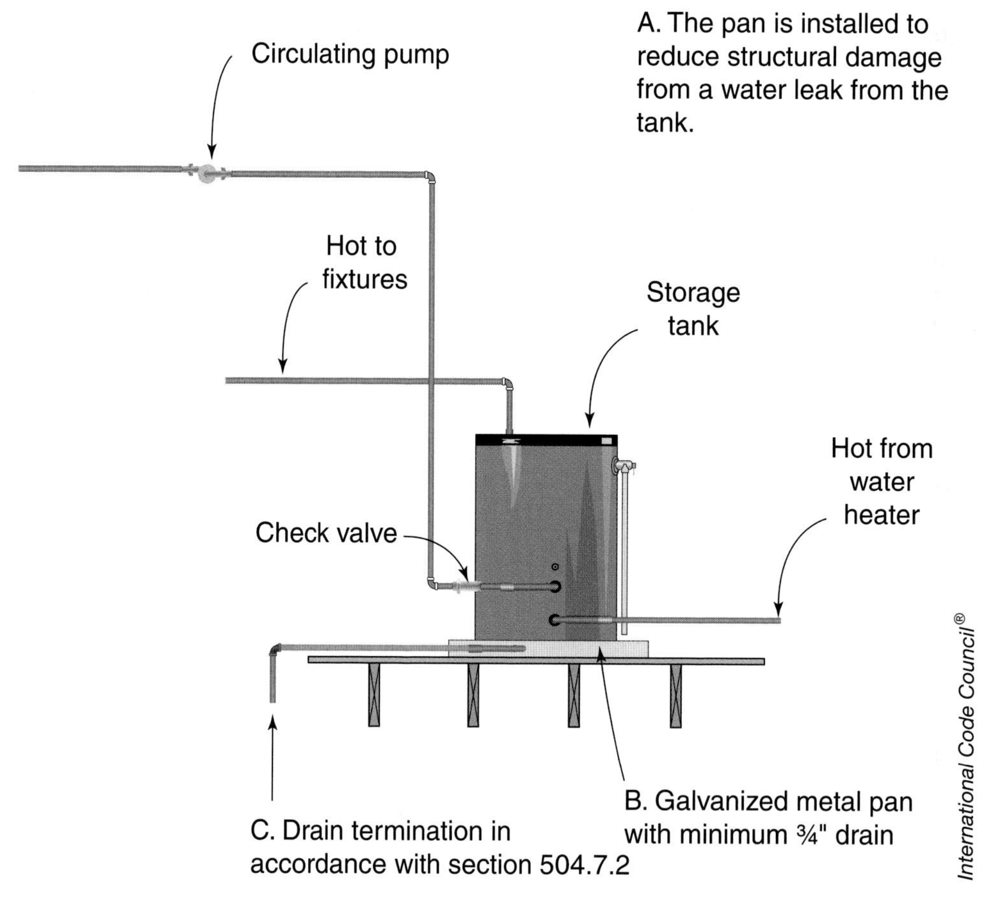

Drain pans are only required for storage-tank-type water heaters or hot water storage tanks.

605

Polyethylene of Raised-Temperature (PE-RT) Plastic Tubing

CHANGE TYPE: Addition

CHANGE SUMMARY: Polyethylene of raised-temperature (PE-RT) plastic hot and cold water tubing and distribution systems are now recognized by the IPC.

2012 CODE:

605 continues

TABLE 308.5 Hanger Spacing

Piping Material	Maximum Horizontal Spacing (Feet)	Maximum Vertical Spacing (Feet)
Polyethylene of Raised Temperature (PE-RT) pipe	2.67 (32 inches)	10[b]

***Reader's Note:** *Other changes may occur in Table 308.5 that will be addressed in different areas of this book; those portions of the table not addressed remain unchanged.*

TABLE 605.3 Water Service Pipe

Material	Standard
Polyethylene of raised temperature (PE-RT) plastic tubing	ASTM F 2769

***Reader's Note:** *Other changes may occur in Table 605.3 that will be addressed in different areas of this book; those portions of the table not addressed remain unchanged.*

TABLE 605.4 Water Distribution Pipe

Material	Standard
Polyethylene of raised temperature (PE-RT) plastic tubing	ASTM F 2769

***Reader's Note:** *Other changes may occur in Table 605.4 that will be addressed in different areas of this book; those portions of the table not addressed remain unchanged.*

TABLE 605.5 Pipe Fittings

Material	Standard
Fittings for polyethylene of raised temperature (PE-RT) plastic tubing	ASSE 1061; ASTM F 877; ASTM F 1807; ASTM F 2080; ASTM F 2098; ASTM F 2159; ASTM F 2434; ASTM F 2735; CSA B137.5
Fittings for cross-linked polyethylene (PEX) plastic tubing	ASSE 1061, ASTM F 877; ASTM F 1807; ASTM F 1960; ASTM F 2080; ASTM F 2003, ASTM F 2159; ASTM F 2434; ASTM F 2735; CSA B137.5

605 continued

***Reader's Note:** *Other changes may occur in Table 605.5 that will be addressed in different areas of this book; those portions of the table not addressed remain unchanged.*

605.25 Polyethylene of Raised Temperature Plastic. Joints between polyethylene of raised temperature plastic tubing and fittings shall be in accordance with Section 605.25.1 and Section 605.25.2.

605.25.1 Flared Joints. Flared pipe ends shall be made by a tool designed for that operation.

605.25.2 Mechanical Joints. Mechanical joints shall be installed in accordance with the manufacturer's instructions. Fittings for polyethylene of raised temperature plastic tubing shall comply with the applicable standards listed in Table 605.5 and shall be installed in accordance with the manufacturer's installation instructions. Polyethylene of raised temperature plastic tubing shall be factory marked with the applicable standards for the fittings that the manufacturer of the tubing specifies for use with the tubing.

Chapter 14:

ASTM

F 2735-09 Standard Specification for SDR9 Cross-linked Polyethylene (PEX) and Raised Temperature (PE-RT) Tubing

F 2769-09 Polyethylene of Raised Temperature (PE-RT) Plastic Hot and Cold-Water Tubing and Distribution Systems

***Reader's Note:** *Other changes may occur in the standards in Chapter 14 that will be identified in different areas of this book.*

CHANGE SIGNIFICANCE: Provisions are now included in the IPC to address the use of polyethylene of raised-temperature (PE-RT) plastic hot and cold water tubing and distribution systems for water service and water distribution. Included in the introduction of PE-RT systems is a new standard, ASTM F 2769, addressing its permitted use. The recognition of PE-RT materials and systems has resulted in additions to several tables within the code. Provisions addressing joints for PE-RT tubing have also been added, as well as a reference to a new fittings standard, ASTM F 2735. The pipe support table has also been amended to provide the maximum support spacing requirements for PE-RT pipe. Taking a closer look at Table 308.5 (Hanger Spacing), it can be seen that the 32-inch limitation is also the standard maximum horizontal spacing required for cross-linked polyethylene (PEX) pipe, cross-linked polyethylene/aluminum/cross-linked polyethylene (PEX-AL-PEX) pipe, polyethylene/aluminum/polyethylene (PE-AL-PE) pipe, and polypropylene (PP) pipe or tubing 1 inch or smaller.

Table 605.3

Polyethylene (PE) Water Service Pipe

CHANGE TYPE: Addition

CHANGE SUMMARY: Reference standard AWWA C901, "Polyethylene (PE) Pressure Pipe and Tubing, ½ in. (13 mm) Through 3 in. (76 mm), for Water Service," has been added to the list of standards in Table 605.3 regulating PE plastic water service pipe and tubing.

2012 CODE:

TABLE 605.3 Water Service Pipe

Material	Standard
Polyethylene (PE) plastic pipe	ASTM D 2239; ASTM D 3035; AWWA C901; CSA-B137.1
Polyethylene (PE) plastic tubing	ASTM D 2737; AWWA C901; CSA B137.1

***Reader's Note:** *Other changes may occur in Table 605.3 that will be addressed in different areas of this book; those portions of the table not addressed remain unchanged.*

Chapter 14:

AWWA

C901-08 Polyethylene (PE) Pressure Pipe and Tubing, ½ in. (13 mm) Through 3 in. (76 mm), for Water Service

***Reader's Note:** *Other changes may occur in the standards in Chapter 14 that will be identified in different areas of this book.*

CHANGE SIGNIFICANCE: AWWA C901 describes polyethylene (PE) pressure pipe and tubing for use primarily as service lines in the construction of underground water distribution systems for use in potable water, reclaimed water, and wastewater service.

AWWA C901 is the AWWA standard for PE pressure pipe, tubing, and fittings, ½ through 3 inches, for water. This standard is primarily for PE water service piping. AWWA C901 can be summarized as follows:

Compounds	2306, 2406, 3406, 3408
Size range	½″ through 3″
Diameter	ID base, OD base, tubing
Pressure classes	80, 100, 125, 160, and 200 lb/in^2
DR	9.0, 11.0, 14.3, 17, and 21

The AWWA C901 provides for a wide selection of compounds and pressure classes. The safety factor is 2, and there is no routine test for each piece of pipe.

Table 605.3

PEX Water Service Pipe

CHANGE TYPE: Addition

CHANGE SUMMARY: Reference standard AWWA C904, "Cross-Linked Polyethylene (PEX) Pressure Pipe, ½ in. (12 mm) Through 3 in. (76 mm) for Water Service," has been added to the list of standards in Table 605.3 regulating PEX water service piping.

2012 CODE:

TABLE 605.3 Water Service Pipe

Material	Standard
Cross-linked polyethylene (PEX) plastic pipe and tubing	ASTM F 876; ASTM F 877; AWWA C904; CSA B137.5

***Reader's Note:** *Other changes may occur in Table 605.3 that will be addressed in different areas of this book; those portions of the table not addressed remain unchanged.*

Chapter 14:

AWWA

C904-06 Cross-Linked Polyethylene (PEX) Pressure Pipe, ½ in. (12 mm) Through 3 in. (76 mm) for Water Service

CHANGE SIGNIFICANCE: The added reference standard AWWA C904 addresses cross-linked polyethylene (PEX) pressure pipe for use primarily as service lines in the construction of underground water distribution systems. This standard describes pipe and tubing made with a materials designation code of PEX 1006 as established in ASTM F 876, "Specification for Cross-Linked Polyethylene (PEX) Tubing." AWWA C904 describes pipe in sizes ½ inch through 3 inches (12 mm through 76 mm) with a standard dimension ratio of 9 (SDR9) and pressure class of 160 psi.

Properly labeled cross-linked polyethylene (PEX) pressure pipe in accordance with the AWWA C904 standard *(Courtesy of Zurn Industries, LLC)*

606.7

Labeling of Water Distribution Pipes in Bundles

CHANGE TYPE: Addition

CHANGE SUMMARY: Water distribution piping that is installed in bundles must now be labeled for content and direction of flow.

2012 CODE: **606.7 Labeling of Water Distribution Pipes in Bundles.** Where water distribution piping is bundled at installation, each pipe in the bundle shall be identified using stenciling or commercially available pipe labels. The identification shall indicate the pipe contents and the direction of flow in the pipe. The interval of the identification markings on the pipe shall not exceed 25 feet. There shall be not less than one identification label on each pipe in each room, space, or story.

CHANGE SIGNIFICANCE: Piping in bundles has been known to cause confusion as to the determination of the contents of the individual pipes. Tracing them back to the source is often difficult and time consuming. Marking the bundled piping with the identification of the contents and the direction of flow will help eliminate the possibility of a cross connection occurring when repairing or renovating this type of plumbing system. The interval of the identification is also crucial in the proper identification of the piping.

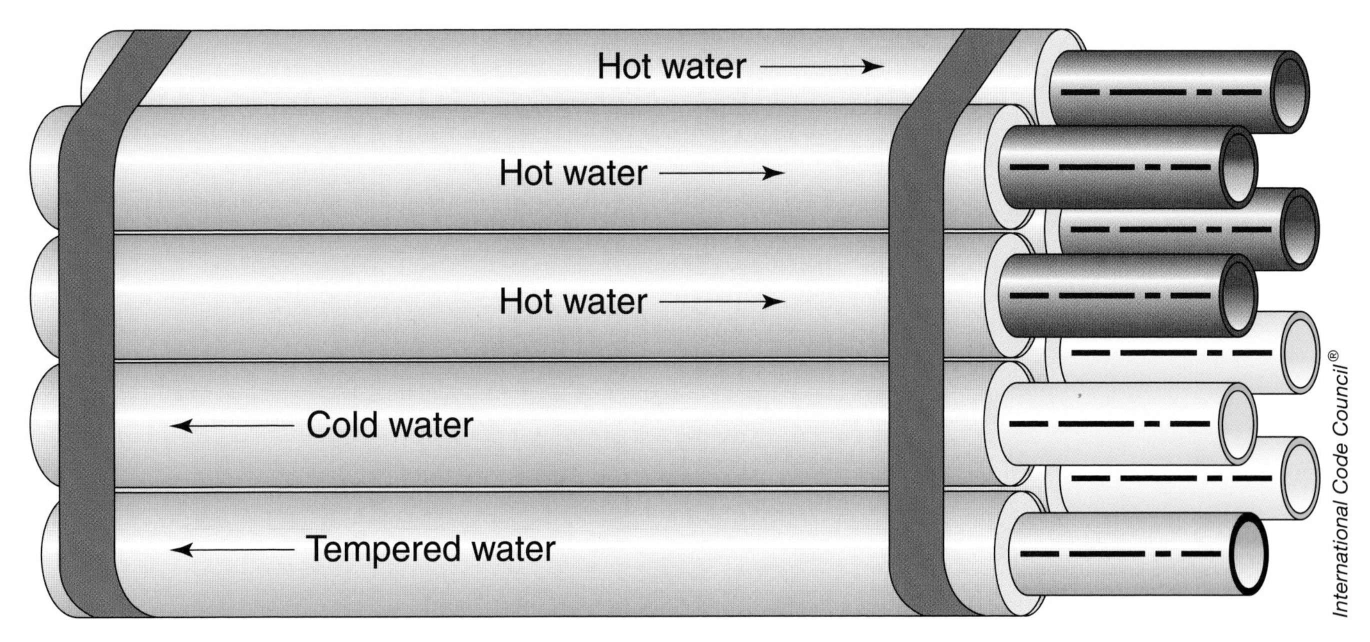

Pipe labeling requirements for bundles

607.1.1

Water-Temperature-Limiting Means

CHANGE TYPE: Modification

CHANGE SUMMARY: A water heater thermostat is now prohibited from being used as the temperature-limiting device where the code requires a limit for hot or tempered water.

2012 CODE: **607.1.1 Temperature Limiting Means.** A thermostat control for a water heater shall not serve as the temperature-limiting means for the purposes of complying with the requirements of this code for maximum allowable hot or tempered water delivery temperatures at fixtures.

CHANGE SIGNIFICANCE: There has been a lot of misguided information in newspaper articles, safety brochures, and Internet websites advising people to turn down the water heater thermostat to prevent scalding. Water heater thermostats cannot be relied upon to control the outlet hot water temperature of a water heater. Although water heater manufacturers have been recommending that installers set thermostats at 120°F to 125°F, and most water heaters are shipped with lower temperature settings, this effort at limiting the temperature has not been sufficient. Plumbing engineers and contractors have continued to recommend that hot water systems be designed with the intended storage temperatures for several reasons. A water heater is sized based on 140°F, so if the temperature setting is turned down, the user will most likely run out of hot water during peak periods. Higher temperatures reduce the threat of *Legionellae* bacteria growth in the water heater tank. Through the use of 140°F water and mixing down to a safe delivery temperature around 120°F to 125°F, a constant hot water delivery temperature is attained. If a water heater burner control thermostat is turned down to a lower temperature, the water heater has a reduced capacity to deliver hot water.

As a result, a water heater thermostat is now prohibited from being used as a temperature-limiting device where the code requires a limit for hot or tempered water.

Typical water heater thermostat *(Courtesy of Bradford White Corporation)*

607.2

Hot or Tempered Water Supply to Fixtures

CHANGE TYPE: Modification

CHANGE SUMMARY: The maximum distance between a hot water supply source and all fixtures served by the supply source has been reduced from 100 feet to 50 feet.

2012 CODE: **607.2 Hot or Tempered Water Supply to Fixtures ~~Supply Temperature Maintenance~~**. ~~Where t~~The developed length of hot or tempered water piping, from the source of hot water to the ~~farthest~~ fixtures that require hot or tempered water, shall not exceed~~s~~ ~~100~~ 50 feet (~~30 480~~ 15240 mm). ~~; the hot water supply system shall be provided with a method of maintaining the temperature in accordance with the *International Energy Code*.~~ Recirculating system piping and heat traced piping shall be considered to be sources of hot or tempered water.

CHANGE SIGNIFICANCE: Fixtures that require hot or tempered water must be located within a specified distance of the hot water source. The previous allowance of 100 feet has been reduced to 50 feet in order to minimize the time it takes to get hot water to a fixture. A large amount of water and energy is wasted unnecessarily while running the water and waiting for the heated water to get to the outlet. Hot water supply is an area where the proper design of the system is critical to create sizeable energy and water savings.

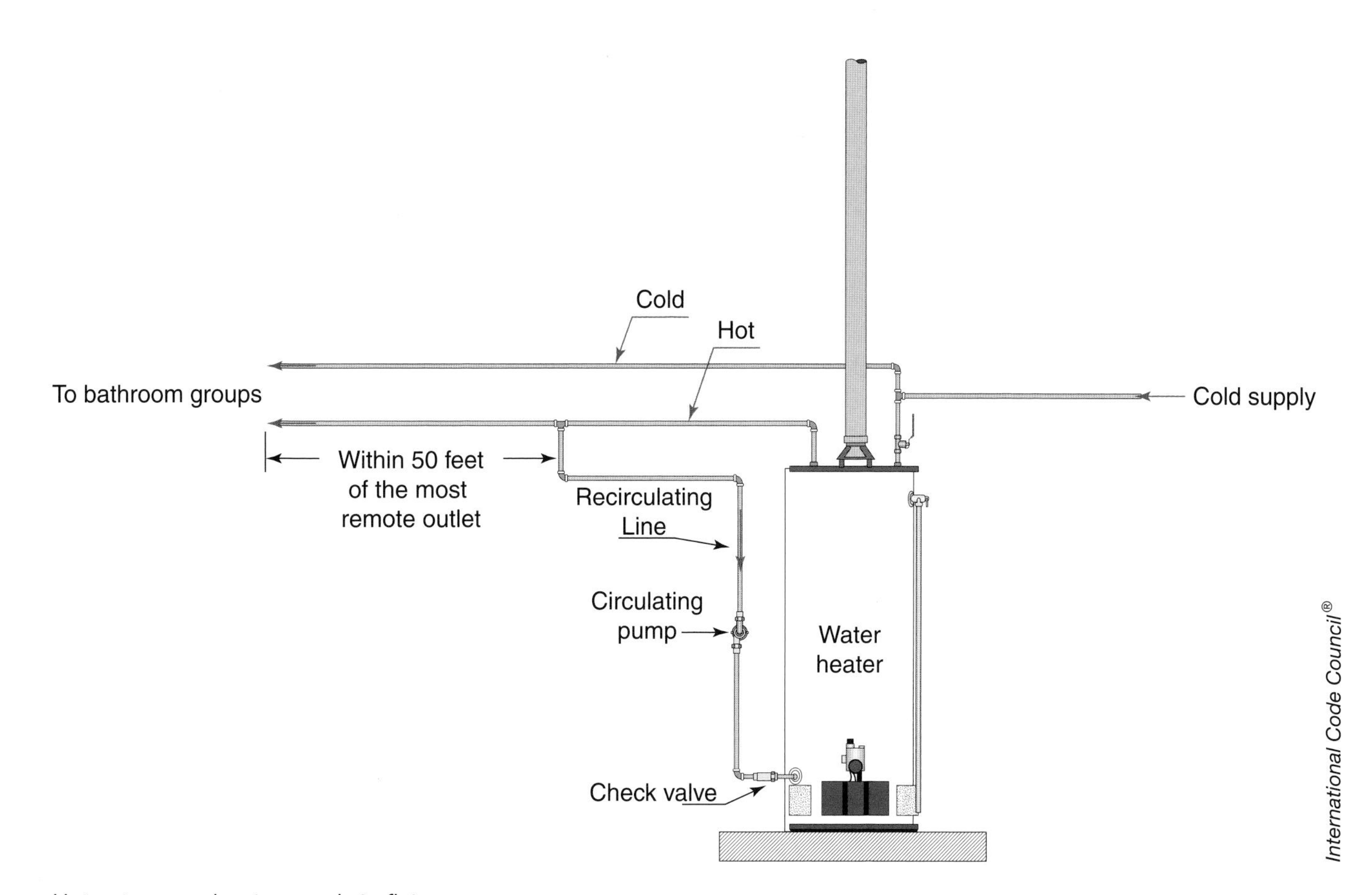

Hot or tempered water supply to fixtures

607.5

Hot Water Piping Insulation

CHANGE TYPE: Addition

CHANGE SUMMARY: The *International Energy Conservation Code* (IECC) requirement for insulating hot water piping in automatic temperature maintenance systems is now also a provision in the IPC.

2012 CODE: **607.5 Pipe Insulation.** Hot water piping in automatic temperature maintenance systems shall be insulated with 1 inch (25 mm) of insulation having a conductivity not exceeding 0.27 Btu per inch/h x ft^2 x °F (1.53 W per 25 mm/m^2 x K). The first 8 feet (2438 mm) of hot water piping from a hot water source that does not have heat traps shall be insulated with 0.5 inch (12.7mm) of material having a conductivity not exceeding 0.27 Btu per inch/h x ft^2 x °F (1.53 W per 25 mm/m^2 x K).

CHANGE SIGNIFICANCE: The minimum insulation requirements for hot water piping in automatic temperature maintenance systems have historically been located in the IECC. Insertion of this important information into the IPC will be a benefit to the user in determining what type and thickness of insulation is required on a hot water piping system. The new text that has been added is a word-for-word extract from the IECC.

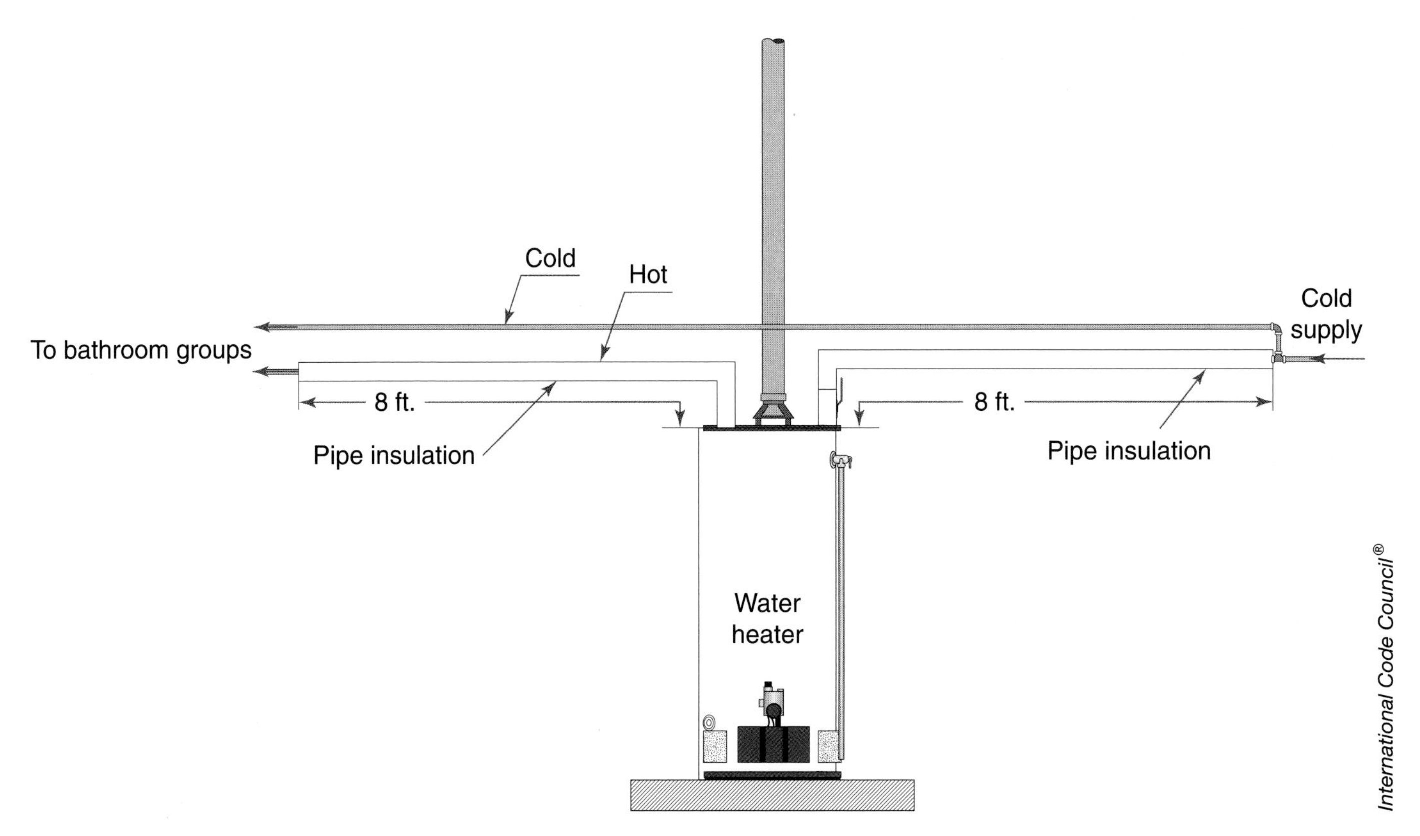

Noncirculating system with no heat traps installed

608.8

Identification of Nonpotable Water

CHANGE TYPE: Modification

CHANGE SUMMARY: Wherever nonpotable water systems are installed, including outside of the building, the piping must be identified.

2012 CODE: 608.8 Identification of Nonpotable Water. ~~In buildings w~~Where nonpotable water systems are installed, the piping conveying the nonpotable water shall be identified either by color marking or metal tags in accordance with Sections 608.8.1 through 608.8.3. All nonpotable water outlets such as hose connections, open-ended pipes, and faucets shall be identified at the point of use for each outlet with the words, "Nonpotable-not safe for drinking." The words shall be indelibly printed on a tag or sign constructed of corrosion-resistant waterproof material or shall be indelibly printed on the fixture. The letters of the words shall be not less than 0.5 inches in height and in colors in contrast to the background on which they are applied.

CHANGE SIGNIFICANCE: The previous code limited the identification of nonpotable water systems to only those systems installed "in buildings" and created confusion as to whether or not outlets not within buildings need to be identified. All systems and outlets for nonpotable water should be identified regardless of location. Expanding the scope of the requirement will result in an increased level of health and welfare protection for the public.

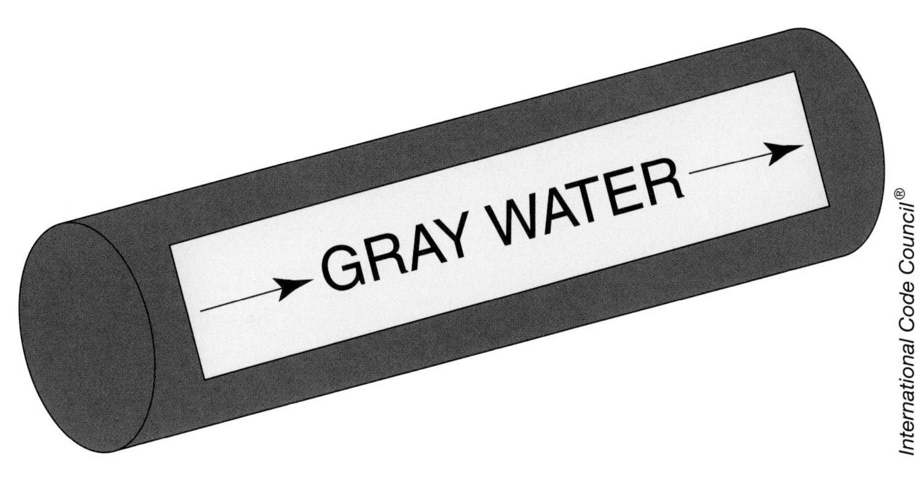

Identification of nonpotable water

704.3, 711.2.1

Horizontal Branch Connections

CHANGE TYPE: Modification

CHANGE SUMMARY: Horizontal branches are now permitted to connect at any point in a stack above or below a horizontal offset. In addition, horizontal branches are now allowed to connect to the base of stacks at a point located not less than 10 times the diameter of the drainage stack downstream from the stack.

2012 CODE: 704.3 Connections to Offsets and Bases of Stacks. Horizontal branches shall connect to the bases of stacks at a point located not less than 10 times the diameter of the drainage stack downstream from the stack. ~~Except as prohibited by Section 711.2, h~~Horizontal branches shall connect to horizontal stack offsets at a point located not less than 10 times the diameter of the drainage stack downstream from the upper stack.

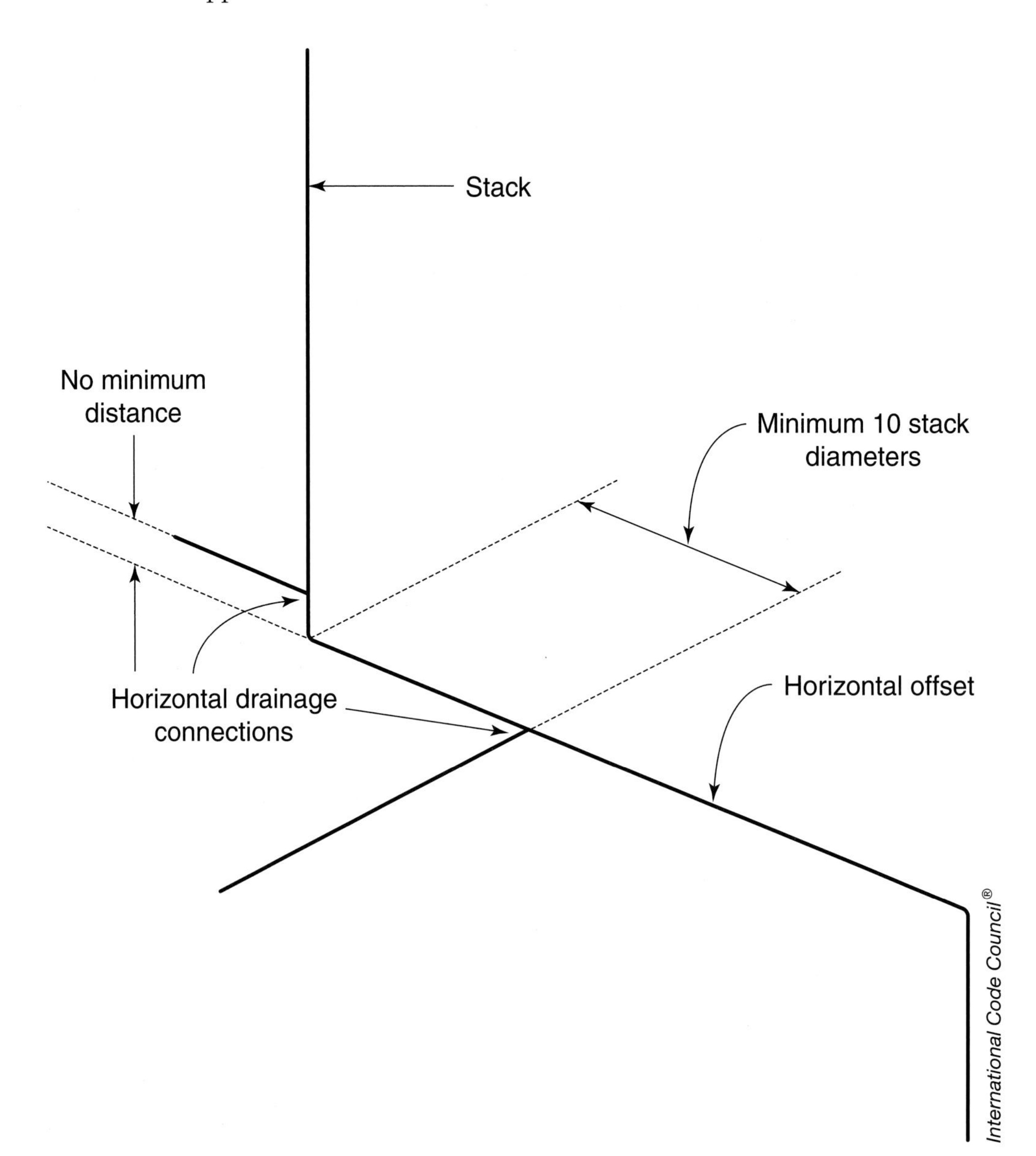

Permitted horizontal branch connection

~~**711.2 Horizontal Branch Connections to Horizontal Stack Offsets.** Where a horizontal stack offset is located more than four branch intervals below the top of the stack, a horizontal branch shall not connect within the horizontal stack offset or within 2 feet (610 mm) above or below such offset.~~

~~711.3.1~~711.2.1 Omission of Vents for Horizontal Stack Offsets. Vents for horizontal stack offsets required by Section 711.3 shall not be required where the stack and its offset are one pipe size larger than required for a building drain [see Table 710.1(1)] and the entire stack and offset are not less in cross-sectional area than that required for a straight stack plus the area of an offset vent as provided for in Section 9~~15~~07. ~~Omission of offset vents in accordance with this section shall not constitute approval of horizontal branch connections within the offset or within 2 feet (610 mm) above or below the offset.~~

CHANGE SIGNIFICANCE: As a result of the deletion of Section 711.2, horizontal branches can now connect at any point in a stack above or below a horizontal offset. Research has shown that the turbulent flow in the horizontal offset occurs within the first 10 pipe (stack) diameters downstream of the stack. This is the same condition that occurs in a building drain, downstream of the base of a stack. At this point the flow in the horizontal pipe becomes nonturbulent open-channel flow. Any connection downstream of where turbulent action is known to occur is now permitted because the concerns for avoiding connections in a turbulent zone are identical. The allowance for horizontal connections to a horizontal offset now mirrors the requirements for connections at the base of the stack.

Table 709.1

Drainage Fixture Units for Bathroom Groups

CHANGE TYPE: Modification

CHANGE SUMMARY: Where fixtures are provided in addition to those in a bathroom group, the footnote addressing additional drainage fixture unit values is now also applicable to those bathroom groups not located within dwelling units.

2012 CODE:

TABLE 709.1 Drainage Fixture Units for Fixtures and Groups

Fixture Type	Drainage Fixture Unit Value as Load Factors	Minimum Size of Trap (inches)
Bathroom group as defined in Section 202 (1.6 gpf water closet)[f]	5	—
Bathroom group as defined in Section 202 (water closet flushing greater than 1.6 gpf)[f]	6	—

f. For fixtures added to a ~~dwelling unit~~ bathroom group, add the dfu value of those additional fixtures to the bathroom group fixture count.

***Reader's Note:** *Other changes may occur in Table 709.1 that will be addressed in different areas of this book; those portions of the table and its footnotes not addressed remain unchanged.*

CHANGE SIGNIFICANCE: In the determination of the drainage fixture unit value where a bathroom group in a dwelling unit includes fixtures in addition to those defined in Section 202 as part of a bathroom group, the value of the additional fixtures is to be added to the bathroom group fixture count. The previous limitation of footnote "f" to only those bathroom groups within dwelling units has now been eliminated because bathroom groups can be located in many different occupancies, not just in dwelling units.

712.3.3

Sump Pump and Ejector Discharge Pipe and Fittings

CHANGE TYPE: Addition

CHANGE SUMMARY: Materials acceptable for use in sump pump and ejector pipe and fittings materials are now specifically listed.

2012 CODE: 712.3.3 Discharge Pip~~inge~~ and Fittings. Discharge pipe and fittings serving sump pumps and ejectors shall be constructed of ~~*approved*~~ materials in accordance with Sections 712.3.3.1 and 712.3.3.2 and shall be *approved.*

712.3.3.1 Materials. Pipe and fitting materials shall be constructed of brass, copper, CPVC, ductile iron, PE, or PVC.

712.3.3.2 Ratings. Pipe and fittings shall be rated for the maximum system operating pressure and temperature. Pipe fitting materials shall be compatible with the pipe material. Where pipe and fittings are buried in the earth, they shall be suitable for burial.

CHANGE SIGNIFICANCE: A simple listing of materials allowed for use in forced main applications for discharge pipe and fittings has added clarity where there was little guidance. It is also now stated that piping and fittings shall be rated for the maximum system operating pressure and temperature, and where buried in the earth they shall be suitable for that purpose.

International Code Council®

PVC is an approved material for sump pump discharge piping.

712.3.5

Sump Pump Connection to the Drainage System

CHANGE TYPE: Modification

CHANGE SUMMARY: Where sump pumps connect to the drainage system, they are now permitted to connect to a building sewer, building drain, soil stack, waste stack, or horizontal branch drain.

2012 CODE:

712.3.5 Pump ~~Ejector~~ Connection to the Drainage System. Pumps connected to the drainage system shall connect to ~~the~~ a *building sewer*, ~~or shall connect to a wye fitting in the~~ *building drain*, soil stack, waste stack, or horizontal branch drain. ~~a minimum of 10 feet (3048 mm) from the base of any soil *stack*, waste *stack* or *fixture drain*.~~ Where the discharge line connects into horizontal drainage piping, the connect~~or~~ion shall be made through a wye fitting into the top of the drainage piping and such wye fitting shall be located not less than 10 pipe diameters from the base of any soil stack, waste stack or fixture drain.

CHANGE SIGNIFICANCE: Building sewers and building drains, soil stacks, waste stacks, and horizontal branch drains are now acceptable points of termination for sump pump discharge lines. When the discharge line connects to a soil stack or waste stack, Section 706.3 would apply,

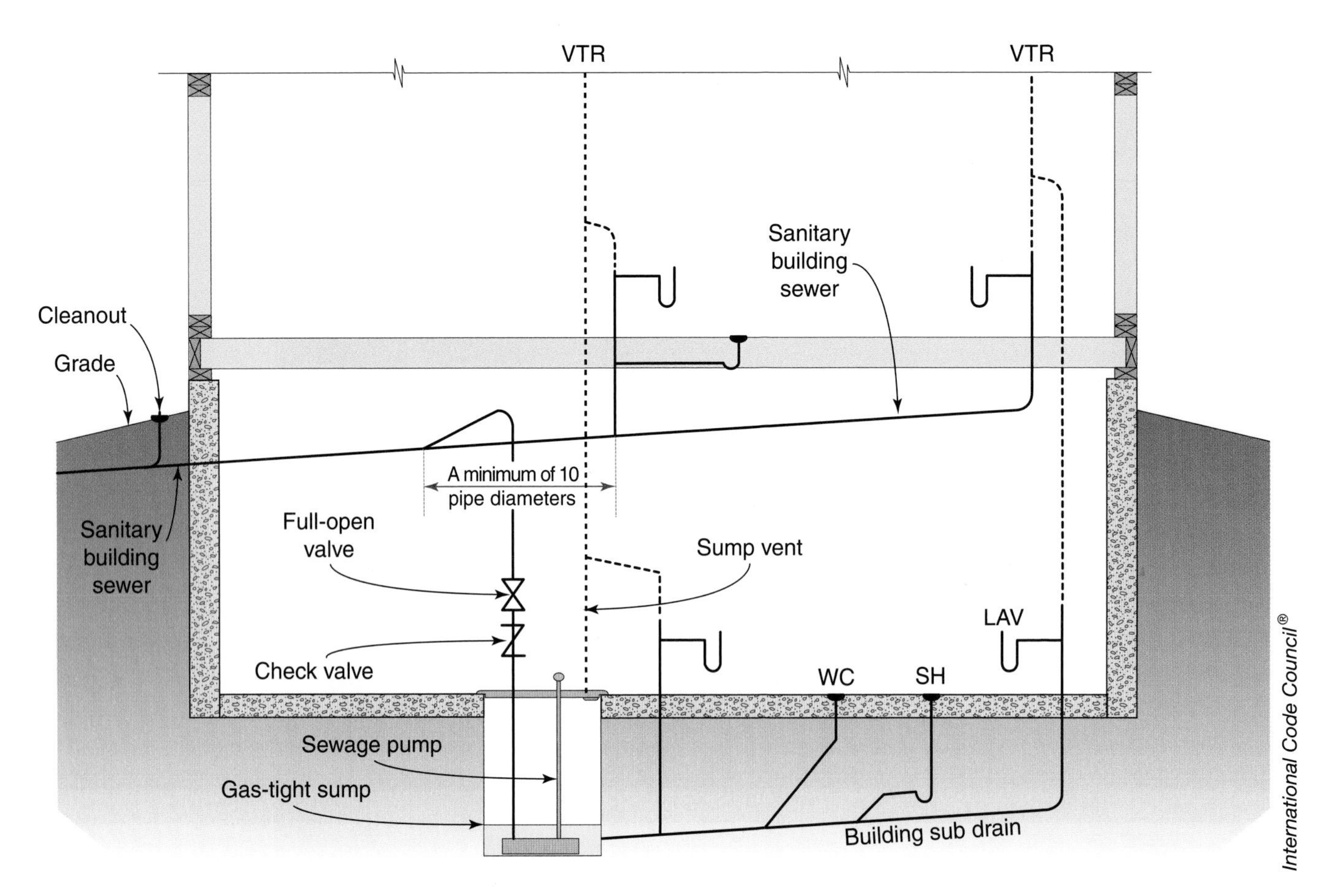

Sump pump connection to drainage system

requiring drainage fittings to be installed to guide sewage and waste in the direction of flow. Fittings acceptable for changes of direction are already addressed in Table 706.3. For a connection to a horizontal drainage pipe, the connection shall continue to be made through a wye-type fitting into the top of the drainage piping, and such wye fitting shall be located not less than 10 pipe diameters from the base of any soil stack waste stack or fixture drain, a change from the previous minimum distance of 10 pipe diameters.

715.1

Fixture Protection from Sewage Backflow

CHANGE TYPE: Modification

CHANGE SUMMARY: In the determination of backwater valve protection from sewage backflow, the use of the finished floor elevation where the fixtures are installed rather than the flood level rim of the fixtures provides a new point of reference.

2012 CODE: **715.1 Sewage Backflow.** Where ~~the flood level rims of~~ plumbing fixtures are <u>installed on a floor with a finished floor elevation</u> below the elevation of the manhole cover of the next upstream manhole in the *public sewer,* such fixtures shall be protected by a backwater valve installed in the *building drain*, or horizontal *branch* serving such fixtures. Plumbing fixtures ~~having flood level rims~~ <u>installed on a floor with a finished floor elevation</u> above the elevation of the manhole cover of the next upstream manhole in the *public sewer* shall not discharge through a backwater valve.

CHANGE SIGNIFICANCE: In the determination of backwater valve protection from sewage backflow, the use of the finished floor elevation where the fixtures are installed provides a new reference point in relationship to the elevation of the next upstream manhole cover. By using the finished floor elevation, a single consistent elevation applies to all fixtures on the floor. A substantial increase in the level of protection for some fixtures, such as lavatories, will be accomplished. Though there will be no real increase in the level of protection for a shower pan, floor drain, or floor mop sink, there will be an overall increase in protection for the structure as a whole. The use of the finished floor elevation will eliminate the reliance on the seal between a floor flange and the wax ring of a water closet to maintain a seal in a pressure situation in the direction of flow for which it is not intended.

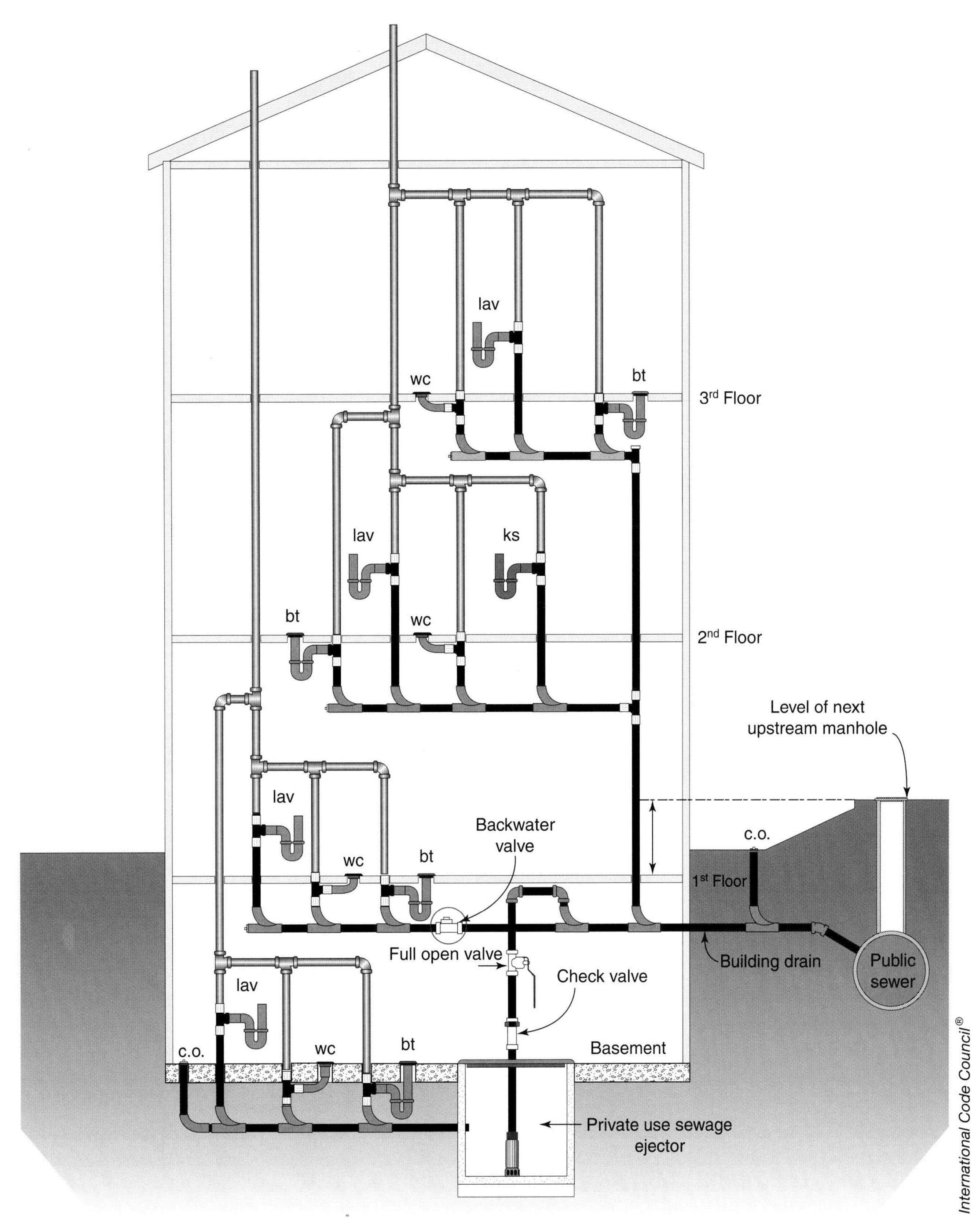

Backwater valve installation

802.1.8

Indirect Discharge of Food Preparation Sinks

CHANGE TYPE: Modification

CHANGE SUMMARY: Sinks used for food preparation and consumption purposes are no longer permitted to connect directly to the drainage system.

2012 CODE: 802.1.8 Food Utensils, Dishes, Pots and Pans Sinks. Sinks used for the washing, rinsing or sanitizing of utensils, dishes, pots, pans or service ware used in the preparation, serving or eating of food shall discharge indirectly through an air gap or an air break ~~or directly connect~~ to the drainage system.

CHANGE SIGNIFICANCE: An air gap or air break is now the only acceptable discharge means for sinks used for the washing, rinsing, or sanitizing of utensils, dishes, pots, pans, or serviceware used in the preparation, serving, or eating of food. Many jurisdictions and health departments believe that having an air gap or air break ensures a higher level of protection than a direct connection. It should be noted that one result of indirect waste discharging into an open receptor is the collection of grease, unless it is properly maintained.

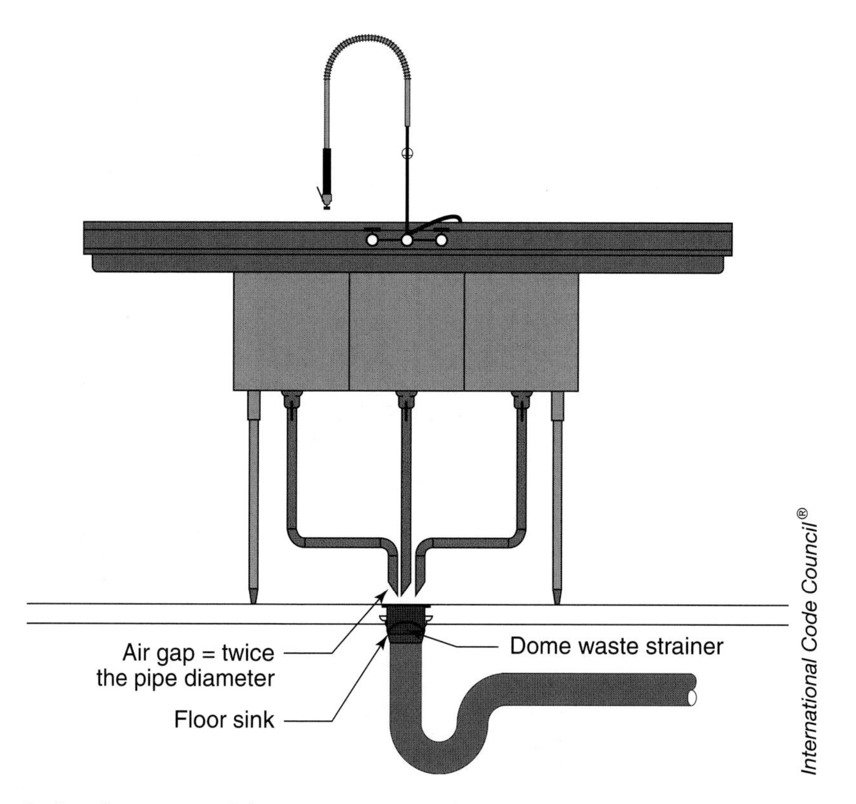

Indirectly connected three-compartment food preparation sink

802.2

Installation of Indirect Waste Piping

CHANGE TYPE: Modification

CHANGE SUMMARY: The thresholds at which indirect waste piping is required to be trapped have been increased and an exception has been added to address clear waste water.

2012 CODE: **802.2 Installation.** All indirect waste piping shall discharge through an air gap or air break into a waste receptor ~~or standpipe~~. Waste receptors and standpipes shall be trapped and vented and shall connect to the building drainage system. All indirect waste piping that exceeds ~~2 feet~~ 30 inches (762 mm) in developed length measured horizontally, or ~~4 feet~~ 54 inches (1372 mm) in total developed length, shall be trapped.

> **Exception:** Where a waste receptor receives only clear water waste and does not directly connect to a sanitary drainage system, the receptor shall not require a trap.

802.2 continues

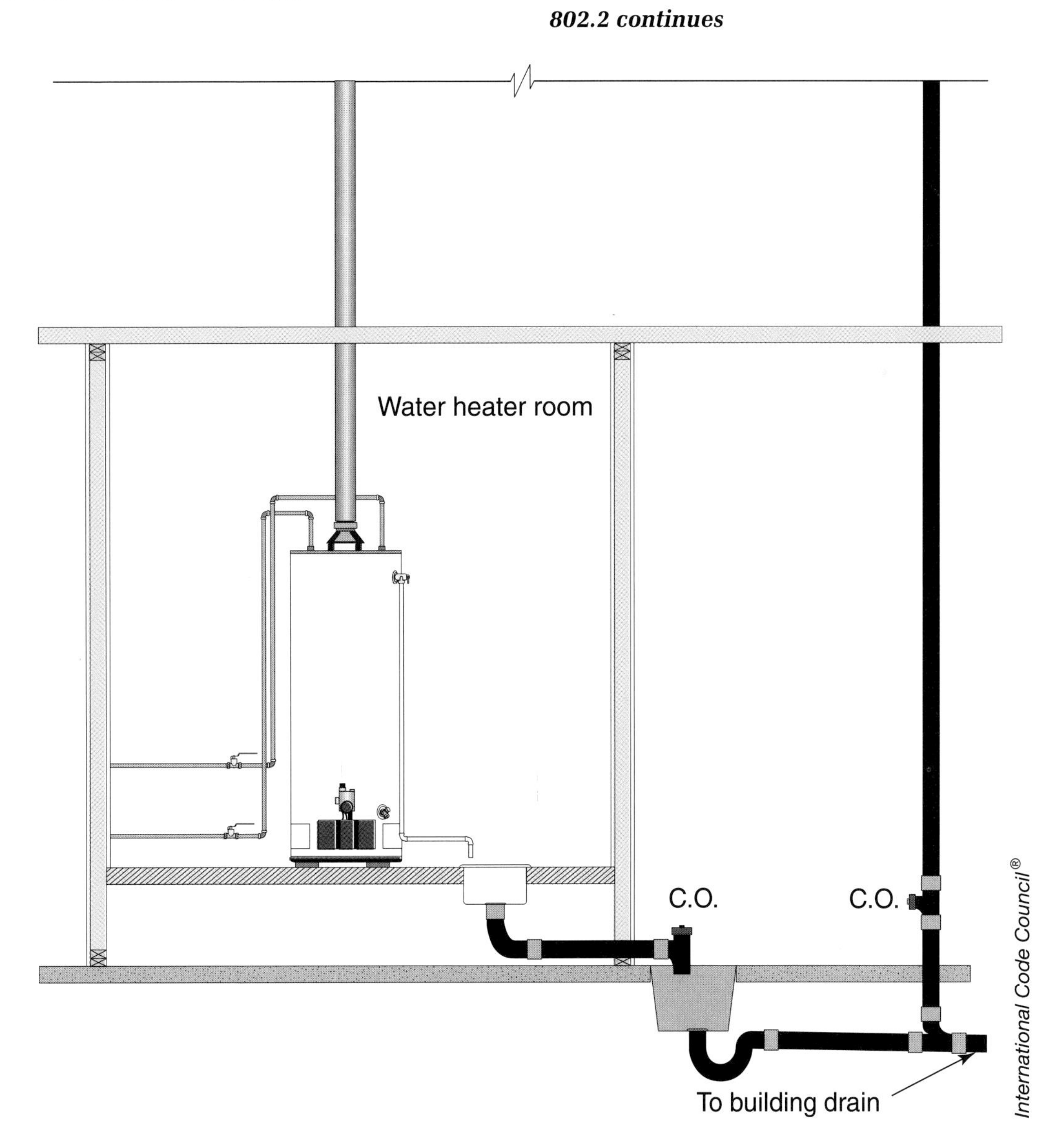

Traps for indirect waste pipes

802.2 continued

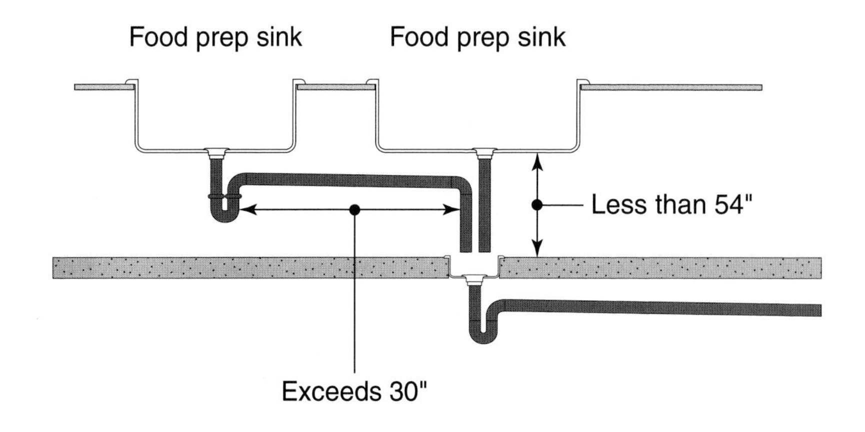

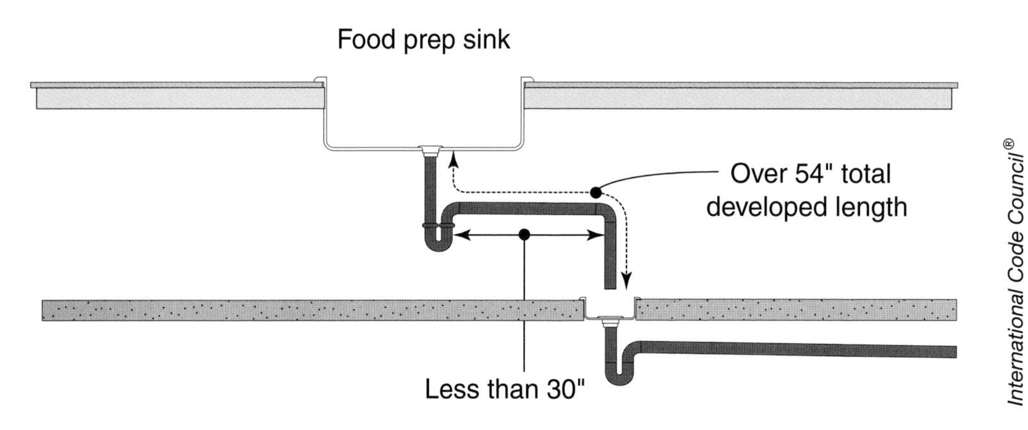

Acceptable means of clear-water discharge to sanitary drainage system

CHANGE SIGNIFICANCE: In the past, indirect waste piping was required to be trapped where it exceeded 24 inches in horizontal developed length or 48 inches in total developed length. The justification for increased developed lengths without a trap, 30 inches for a horizontal measurement and 54 inches in total developed length, is based on Section 1002.1 addressing fixture traps, including the allowance of 30 inches center-to-center for a combination fixture as permitted in Exception 2. The 54-inch total developed length allowance is simply the 30-inch horizontal length allowance plus the 24-inch vertical distance allowed from a fixture to its trap. The changes are considered logical and provide consistency with other allowances in the code. The new exception is fundamental in that traps are unnecessary for clear-water waste in an indirect piping system.

802.3

Prohibited Locations for Waste Receptors

CHANGE TYPE: Modification

CHANGE SUMMARY: The list of locations where waste receptors cannot be located has been expanded to specifically include plenums, crawlspaces, attics, and interstitial spaces above ceilings and below floors.

2012 CODE: 802.3 Waste Receptors. Every waste receptor shall be of an approved type. A removable strainer or basket shall cover the waste outlet of waste receptors. Waste receptors shall be installed in ventilated spaces. Waste receptors shall not be installed in bathrooms, ~~or~~ toilet rooms, plenums, crawl spaces, attics, interstitial spaces above ceilings and below floors, or in any inaccessible or unventilated space such as a closet or storeroom. Ready access shall be provided to waste receptors.

CHANGE SIGNIFICANCE: An open unattended trap located in a plenum, crawlspace, attic, or interstitial space above a ceiling or below a floor is an opportunity for disaster that could result in major damage to the

802.3 continues

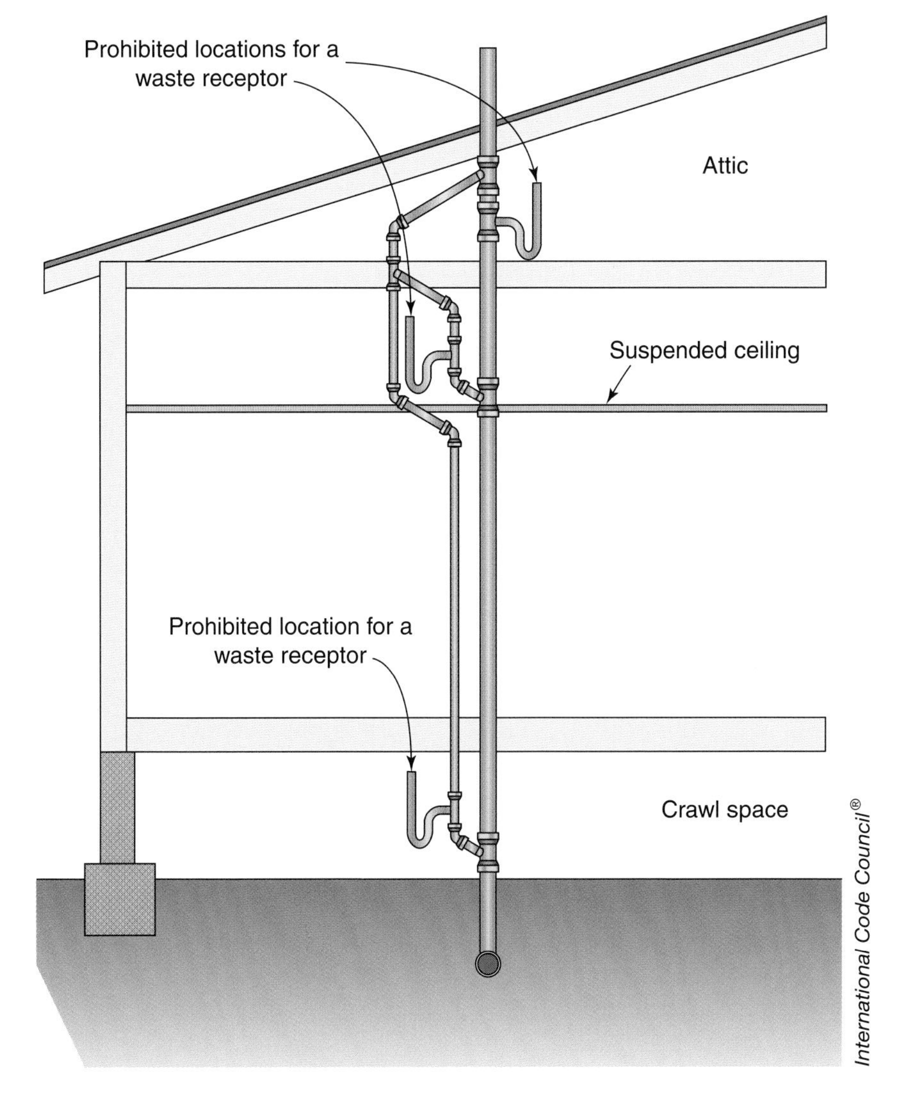

Prohibited locations for waste receptors

802.3 continued building structure. The discharge of condensate drains in these locations is a common practice and has been the cause of much damage. Because these fixtures often go unnoticed for extended periods of time, they are prone to flooding, resulting in structural damage and severe unsanitary conditions. Trap primers provide no assurance that the seals in the traps will be maintained. Trap primers are mechanical devices that are subject to failure and are often not repaired but only isolated from the system. Traps in attics are also subject to freezing, which increases the chance of failure.

901.3, 918.8

Air Admittance Valves for Chemical Waste Vent Systems

CHANGE TYPE: Modification

CHANGE SUMMARY: Air admittance valves complying with reference standard ASSE 1049, "Performance Requirements for Individual and Branch-Type Air Admittance Valves for Chemical Waste Systems," are now permitted to be used for venting chemical waste systems.

2012 CODE: 901.3 Chemical Waste Vent Systems. The vent system for a chemical waste system shall be independent of the sanitary vent system and shall terminate separately through the roof to the ~~open air~~ outdoors or to an air admittance valve that complies with ASSE 1049. Air admittance valves for chemical waste systems shall be constructed of materials approved in accordance with Section 702.5 and shall be tested for chemical resistance in accordance with ASTM F 1412.

~~917.8~~ 918.8 Prohibited Installations. Air admittance valves shall not be installed in non-neutralized special waste systems as described in Chapter 8 except where such valves are in compliance with ASSE 1049, are constructed of materials approved in accordance with Section 702.5 and are tested for chemical resistance in accordance with ASTM F 1412.

Air admittance valves shall not be located in spaces utilized as supply or return air plenums.

Chapter 14:

ASSE

1049-2009 Performance Requirements for Individual and Branch Type Air Admittances Valves for Chemical Waste Systems.

ASTM

F 1412-01 Standard Specification for Polyolefin Pipe and Fittings for Corrosive Waste Drainage Systems.

Studor ChemVent air admittance valve *(Courtesy of IPS Corporation)*

CHANGE SIGNIFICANCE: The American Society of Sanitary Engineers (ASSE) has recently developed ANSI/ASSE Standard 1049, "Performance Requirements for Individual and Branch Type Air Admittance Valves for Chemical Waste Systems." Air admittance valves (AAVs) that are in compliance with ANSI/ASSE 1049, meet the materials requirements of Section 702.5, and are tested to ASTM F 1412 for chemical resistance are now allowed to serve as the vent for non-neutralized special waste systems as an alternative to chemical waste vent piping terminating outdoors. It is quite common to see laboratory sinks that receive chemical waste located in islands in the middle of rooms. To vent the traps for these sinks using vent piping that can only terminate outdoors requires extensive labor and material. Allowing the use of the AAVs will significantly reduce the cost of plumbing these laboratories and similar facilities.

903.5

Location of Vent Terminals

CHANGE TYPE: Modification

CHANGE SUMMARY: The prohibited locations for vent terminals have been revised to provide consistency with the IMC.

2012 CODE: **~~904.5~~ 903.5 Location of Vent Terminal.** An open vent terminal from a drainage system shall not be located directly beneath any door, openable window, or other air intake opening of the building or of an adjacent building, and any such vent terminal shall not be within 10 feet (3048 mm) horizontally of such an opening unless it is at least ~~2 feet (610 mm)~~ 3 feet (914 mm) above the top of such opening.

CHANGE SIGNIFICANCE: Previous provisions required that a vent terminal located within 10 feet horizontally of a door, openable window, or other air intake opening be at least 2 feet above the top of the opening. However, the vertical separation requirement of 2 feet was inconsistent with similar sections in the other International Codes, such as *International Mechanical Code* (IMC) Section 401.4 #3; *International Residential Code* (IRC) Section G2427.6.6; IRC Section G2427.8 #1; *International Fuel Gas Code* (IFGC) Section 503.6.7; IFGC Section 503.8 #1; and IFGC Section 618.5 #1. The 3-foot dimension has been around for many years and has been shown to be very effective in preventing sources of contamination from making their way into building openings.

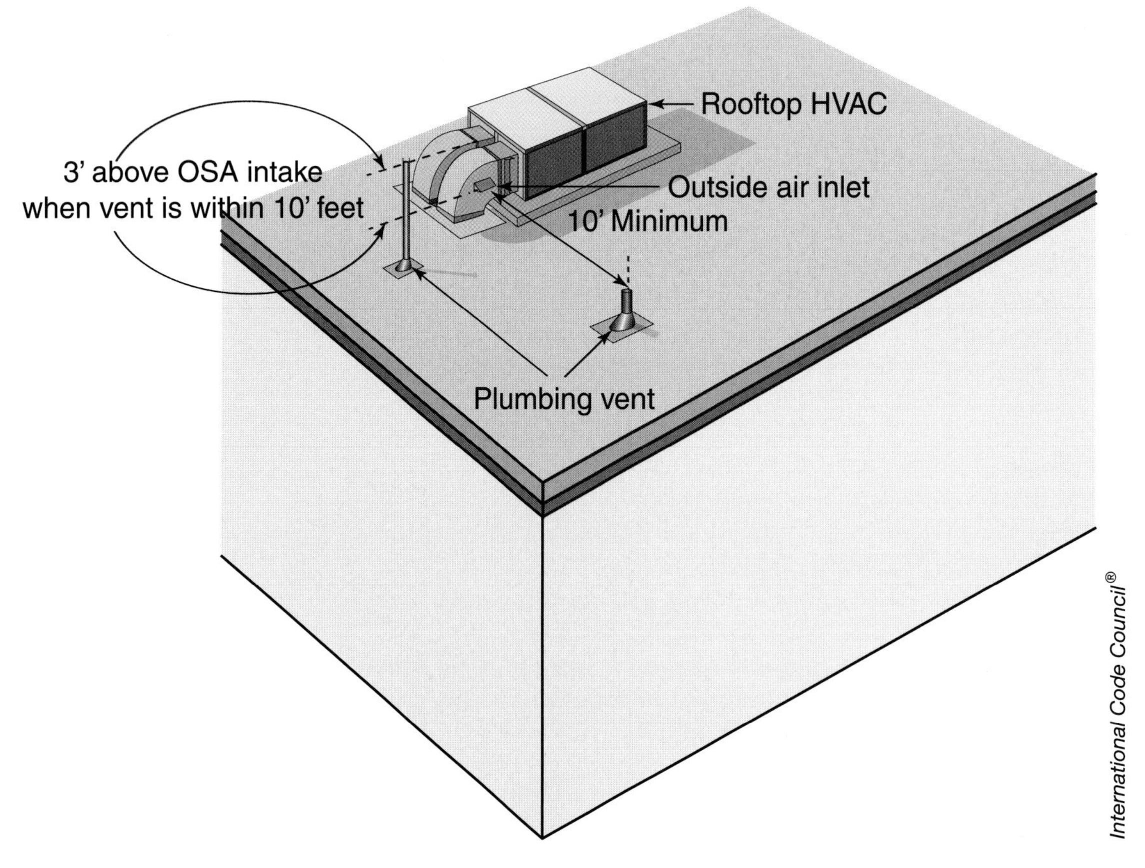

Approved location of vent terminals

915.2

Combination Waste and Vent System Sizing

CHANGE TYPE: Clarification

CHANGE SUMMARY: The length of a combination waste and vent system is unlimited.

2012 CODE: **~~912.2.2~~ 915.2 Connection.** The combination ~~drain~~ waste and vent system shall be provided with a dry vent connected at any point within the system or the system shall connect to a horizontal drain that is vented in accordance with one of the venting methods specified in this chapter. Combination ~~drain~~ waste and vent systems connecting to building drains receiving only the discharge from a stack or stacks shall be provided with a dry vent. The vent connection to the combination ~~drain~~ waste and vent pipe shall extend vertically to a ~~point not less than~~ of 6 inches (152 mm) above the flood level rim of the highest fixture being vented before offsetting horizontally. The horizontal length of a combination waste and vent system shall be unlimited.

CHANGE SIGNIFICANCE: The change now clarifies the intent of the code by adding that a combination waste and vent system sized per the code is unlimited in horizontal length. There is no technical reason to continue to restrict the horizontal length of these piping systems that are commonly used where floor drains are installed in large, open areas that cannot accommodate vertical vent risers from the floor drains. These systems are practical to use in large commercial kitchens, markets, warehouses, and where island sinks are installed in laboratories and classrooms.

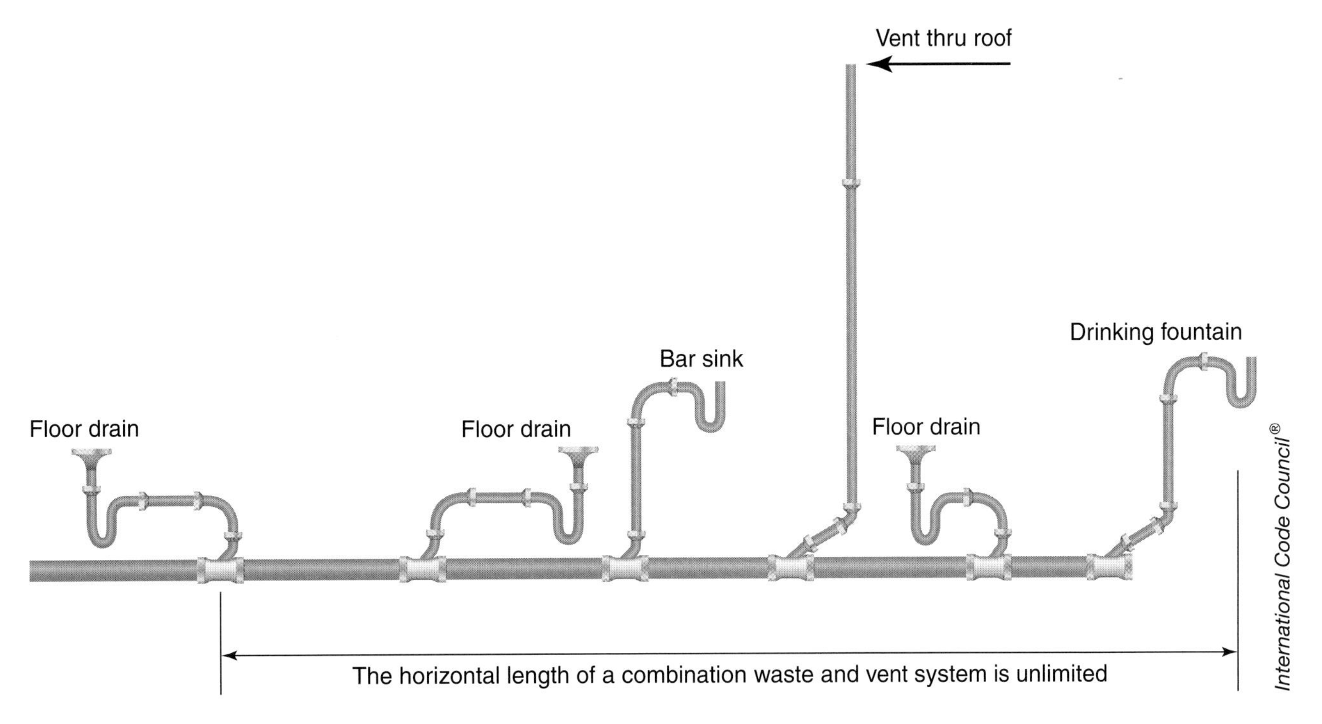

Combination waste and vent system

917

Single-Stack Vent Systems

CHANGE TYPE: Addition

CHANGE SUMMARY: The single-stack vent system method, similar to the Philadelphia stack drainage system, has been added as an acceptable venting system.

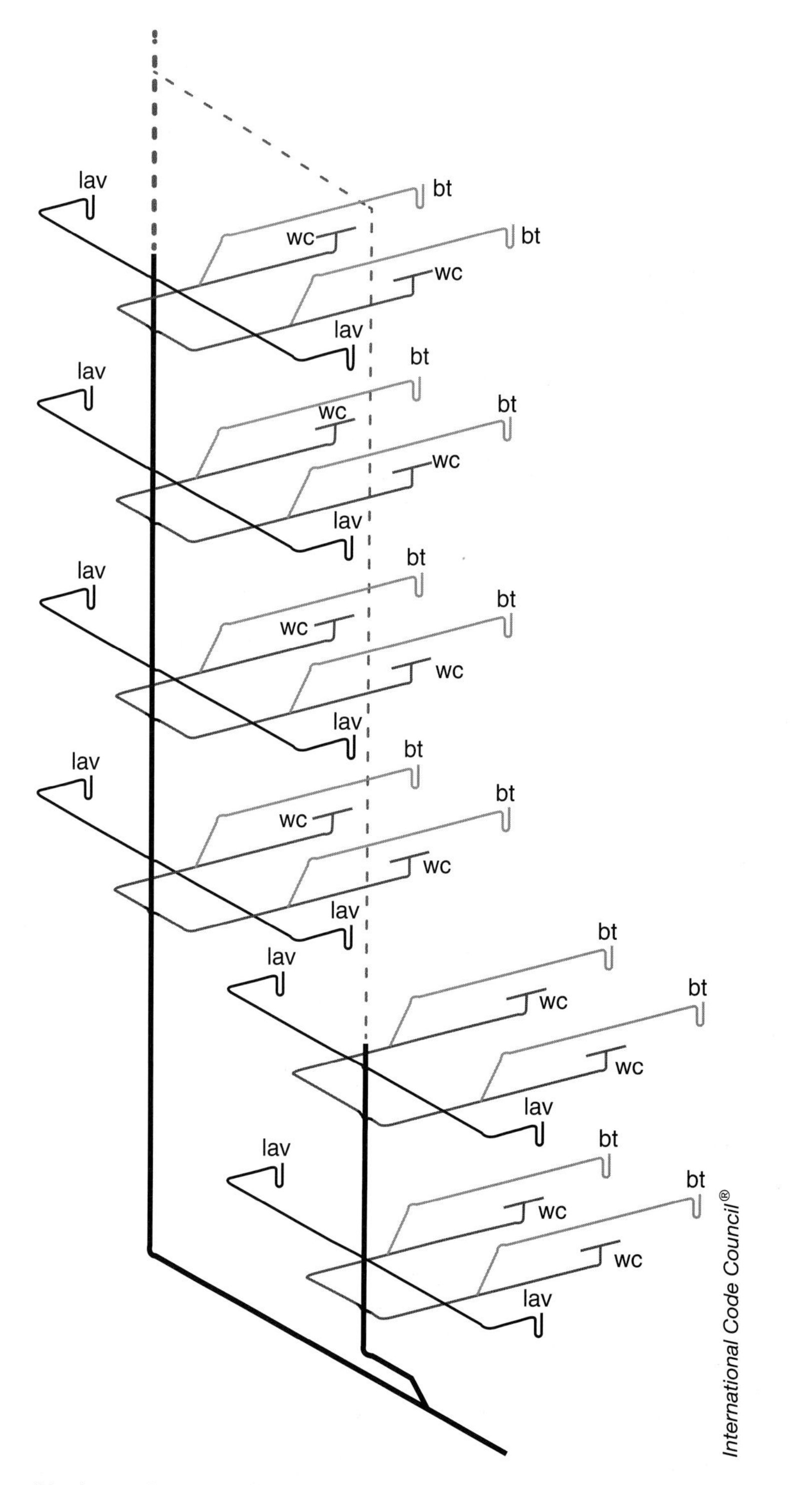

Single-stack system for a six-story building

2012 CODE:

SECTION 917
SINGLE STACK VENT SYSTEM

917.1 Where Permitted. A drainage stack shall serve as a single stack vent system where sized and installed in accordance with Sections 917.2 through 917.9. The drainage stack and branch piping shall be the vents for the drainage system. The drainage stack shall have a stack vent.

917.2 Stack Size. Drainage stacks shall be sized in accordance with Table 917.2. Stacks shall be uniformly sized based on the total connected drainage fixture unit load. The stack vent shall be the same size as the drainage stack. A 3-inch stack shall serve not more than two water closets.

917.3 Branch Size. Horizontal branches connecting to a single stack vent system shall be sized in accordance with Table 710.1(2). Not more than one water closet shall discharge into a 3 inch (76.2 mm) horizontal branch at a point within a developed length of 18 inches (457.2 mm) measured horizontally from the stack.

Where a water closet is within 18 inches (457.2 mm) measured horizontally from the stack and not more than one fixture with a drain size of not more than 1-1/2 inch (38.1 mm) connects to a 3 inch (76.2 mm) horizontal branch, the branch drain connection to the stack shall be made with a sanitary tee.

917.4 Length of Horizontal Branches. The length of horizontal branches shall conform to the requirements of Sections 917.4.1 through 917.4.3.

917.4.1 Water Closet Connection. Water closet connections shall be not greater than 4 feet (1219 mm) in developed length measured horizontally from the stack.

917 continues

TABLE 917.2 Single Stack Size

	MAXIMUM CONNECTED DRAINAGE FIXTURE UNITS		
Stack Size (inches)	Stacks less than 75 feet in height	Stacks 75 feet to less than 160 feet in height	Stacks 160 feet and greater in height
3	24	NP	NP
4	225	24	NP
5	480	225	24
6	1,015	480	225
8	2,320	1,015	480
10	4,500	2,320	1,015
12	8,100	4,500	2,320
15	13,600	8,100	4,500

For SI: 1 inch = 25.4 mm, 1 foot = 304.8 mm.

917 continued

Exception: Where the connection is made with a sanitary tee, the maximum developed length shall be 8 feet (2438 mm).

917.4.2 Fixture Connections. Fixtures other than water closets shall be located not greater than 12 feet (3657 mm) in developed length measured horizontally from the stack.

917.4.3 Vertical Piping in Branch. The length of vertical piping in a fixture drain connecting to a horizontal branch shall not be considered in computing the fixture's distance in developed length measured horizontally from the stack.

917.5 Minimum Vertical Piping Size from Fixture. The vertical portion of piping in a fixture drain to a horizontal branch shall be 2 inches (50.8 mm). The minimum size of the vertical portion of piping for a water supplied urinal or standpipe shall be 3 inches (76.2 mm). The maximum vertical drop shall be 4 feet. Fixture drains that are not increased in size, or have a vertical drop in excess of 4 feet shall be individually vented.

917.6 Additional Venting Required. Additional venting shall be provided where more than one water closet discharges to a horizontal branch and where the distance from a fixture trap to the stack exceeds the limits in Section 917.4. Where additional venting is required, the fixture(s) shall be vented by individual vents, common vents, wet vents, circuit vents, or a combination waste and vent pipe. The dry vent extensions for the additional venting shall connect to a branch vent, vent stack, stack vent, air admittance valve, or shall terminate outdoors.

917.7 Stack Offsets. Where fixture drains are not connected below a horizontal offset in a stack, a horizontal offset shall not be required to be vented. Where horizontal branches or fixture drains are connected below a horizontal offset in a stack, the offset shall be vented in accordance with Section 907. Fixture connections shall not be made to a stack within 2 feet (609.6 mm) above or below a horizontal offset.

917.8 Prohibited Lower Connections. Stacks greater than 2 branch intervals in height shall not receive the discharge of horizontal branches on the lower two floors. There shall be no connections to the stack between the lower two floors and a distance of not less than 10 pipe diameters downstream from the base of the single stack vented system.

917.9 Sizing Building Drains and Sewers. The building drain and building sewer receiving the discharge of a single stack vent system shall be sized in accordance with Table 710.1(1).

CHANGE SIGNIFICANCE: In a *single-stack vent system,* the drainage stack serves as both a single-stack drainage and vent system. The drainage stack and branch piping are considered as vents for the drainage system as a whole. Pipe sizing in a single-stack drainage system is larger than in a conventional one; however, a significant cost saving is achieved by the reduction in the vent piping needed. This venting system serves as a viable alternative to the more traditional systems that are being used. The length of trap arms is limited and the vertical drop from the fixtures is also oversized. Fixture connections that do not meet the requirement for a single-stack system must be conventionally vented. Oversized, unvented S-traps are the major components of the one-pipe system instead of the conventionally sized and vented P-traps and fixtures that allow water to run off after the tap is closed to fill the traps with water to maintain the trap seal. The length of the trap arm is limited to reduce any suction buildup, and the stack is oversized to limit the internal air pressure and vacuum buildup. Stacks greater than two branch intervals in height are prohibited from receiving the discharge from horizontal branches on the lower two floors. The separate stack serving the lower two floors is required to connect to the building drain at a distance of not less than 10 pipe diameters downstream from the base of the connection of the single-stack vented system. This proven method has been laboratory tested to determine sizing and installation requirements that provide proper venting to the drainage system.

1002.1

Floor Drains in Multi-Level Parking Structures

CHANGE TYPE: Modification

CHANGE SUMMARY: Floor drains in multi-level parking garages are no longer required to have individual traps, provided the drains are connected to a main trap before discharge to a combined sewer.

2012 CODE: **1002.1 Fixture Traps.** Each plumbing fixture shall be separately trapped by a liquid-seal trap, except as otherwise permitted by this code. The vertical distance from the fixture outlet to the trap weir shall not exceed 24 inches (610 mm), and the horizontal distance shall not exceed 30 inches (762 mm) measured from the centerline of the fixture outlet to the centerline of the inlet of the trap. The height of a clothes washer standpipe above a trap shall conform to Section 802.4. A fixture shall not be double trapped.

Exceptions:

1.–3. (no changes to text)

4. Where floor drains in multi-level parking structures are required to discharge to a combined building sewer system, the floor drains shall not be required to be individually trapped provided that they are connected to a main trap in accordance with Section 1103.1.

CHANGE SIGNIFICANCE: Floor drain traps located in unheated multi-level parking structures are problematic because the traps can be damaged during freezing conditions. The liquid in traps for floor drains in covered parking levels usually evaporates, as there is little, if any, water runoff on the covered levels. Heat tracing and insulation are also not reliable in these locations. A main trap for the parking structure floor drain system is required in accordance with Section 1103.1. The trap requirement will prevent sewer gas from entering the multi-level parking structure.

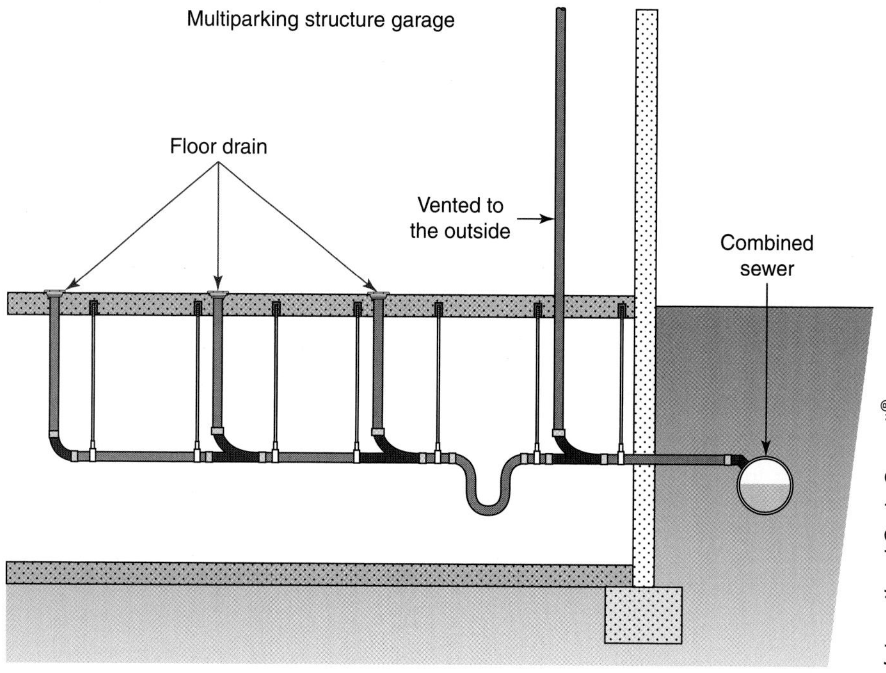

Exemption for floor drain traps in parking garages

1003.1

Interceptors and Separators

CHANGE TYPE: Clarification

CHANGE SUMMARY: It has been clarified that required interceptors and separators are permitted to be located downstream of the building drain.

2012 CODE: 1003.1 Where Required. Interceptors and separators shall be provided to prevent the discharge of oil, grease, sand and other substances harmful or hazardous to the ~~building drainage system, the~~ public sewer, the private sewage disposal system, or the sewage treatment plant or processes.

CHANGE SIGNIFICANCE: The IPC now recognizes that waste must travel through some portion of the building drainage system in order to get to the interceptor. Removing the text "building drainage system" has addressed the misconception that a device is required to be installed adjacent to each and every fixture that discharges any liquid that may need to be separated prior to entering the public or private systems.

International Code Council®

The protection of public sewers

1003.3.1

Alternate Grease Interceptor Locations

CHANGE TYPE: Modification

CHANGE SUMMARY: Grease interceptors are now permitted to be installed in series instead of requiring replacement of an existing grease interceptor that is too small.

2012 CODE: **1003.3.1 Grease Interceptors and Automatic Grease Removal Devices Required.** A grease interceptor or automatic grease removal device shall be required to receive the drainage from fixtures and equipment with grease-laden waste located in food preparation areas, such as in restaurants, hotel kitchens, hospitals, school kitchens, bars, factory cafeterias and clubs. Fixtures and equipment shall include pot sinks, prerinse sinks; soup kettles or similar devices; wok stations; floor drains or sinks into which kettles are drained; automatic hood wash units and dishwashers without prerinse sinks. Grease interceptors and automatic grease removal devices shall receive waste only from fixtures and equipment that allow fats, oils or grease to be discharged. Where lack of space or other constraints prevent the installation or replacement of a grease interceptor, one or more grease interceptors shall be permitted to be installed on or above the floor and upstream of an existing grease interceptor.

CHANGE SIGNIFICANCE: Combinations of grease interceptors are now allowed to be provided for renovation projects involving existing buildings where there is insufficient space or it is cost prohibitive to install a large enough in-ground interceptor (usually a gravity type) to meet local sewer ordinance requirements. Ordinances have become more restrictive to address the need to prevent public sewer lines from excess amounts of grease and lessen the burden on the sewage treatment facilities.

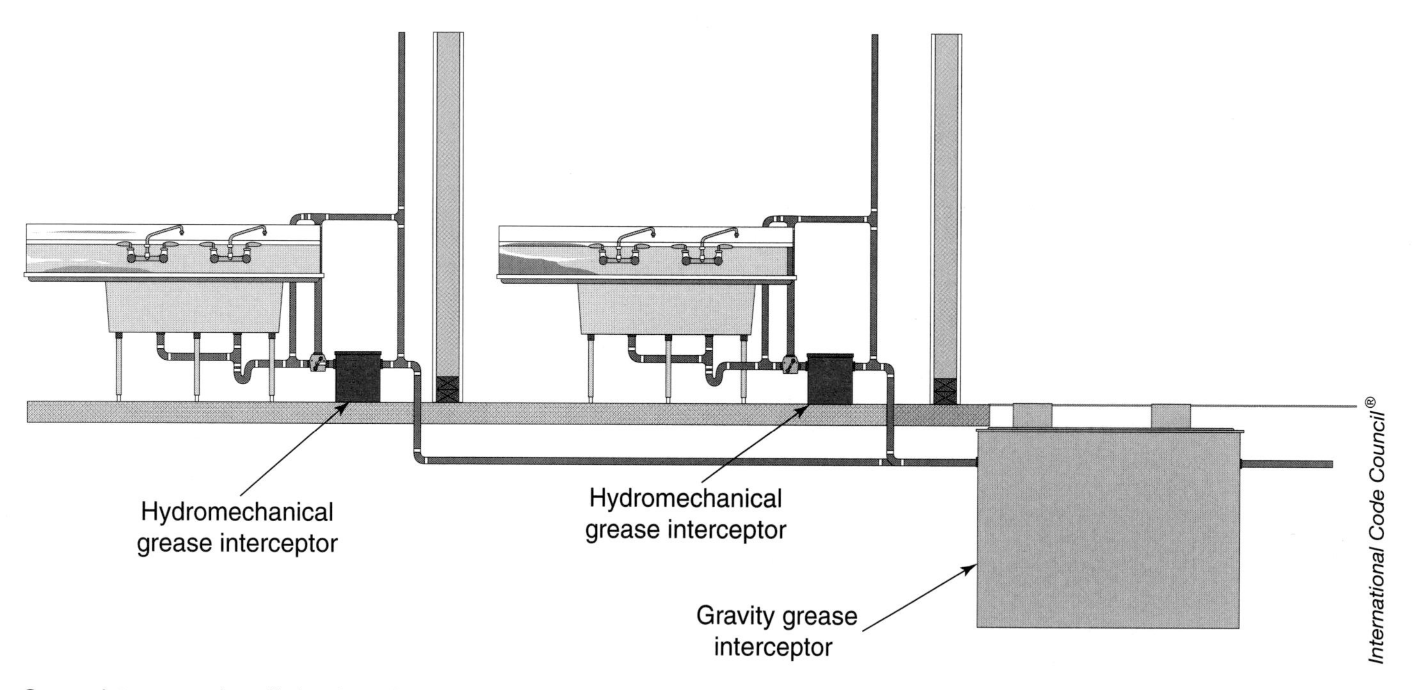

Grease interceptor installation in series

1003.3.4

Hydromechanical Grease Interceptors

CHANGE TYPE: Clarification

CHANGE SUMMARY: In regard to grease interceptors, the new term "hydromechanical" provides a clear distinction from gravity interceptors in order to provide clarity regarding the applicable requirements for each type of interceptor.

2012 CODE: 1003.3.4 Hydromechanical Grease Interceptors and Automatic Grease Removal Devices. Hydromechanical grease interceptors and automatic grease removal devices shall be sized in accordance with ~~PDI G101,~~ ASME.A112.14.3 Appendix A, ASME A112.14.4, CSA B481.3, or PDI G101. Hydromechanical grease interceptors and automatic grease removal devices shall be designed and tested in accordance with ~~PDI G101~~, ASME A112.14.3 Appendix A, ~~or~~ ASME A112.14.4, CSA B481.1, PDI G101, or PDI G102. Hydromechanical grease interceptors and automatic grease removal devices shall be installed in accordance with the manufacturer's instructions. Where manufacturer's instructions are not provided, hydromechanical grease interceptors and grease removal devices shall be installed in compliance with ASME A112.14.3, ASME A112.14.4, CSA B481.3 or PDI G101. This section shall not apply to gravity grease interceptors.

> **Exception:** ~~Interceptors that have a volume of not less than 500 gallons (1893 L) and that are located outdoors shall not be required to meet the requirements of this section.~~

Chapter 14:

PDI

> G102 Testing and Certification for Grease Interceptors with FOG Sensing and Alarm Devices

1003.3.4 continues

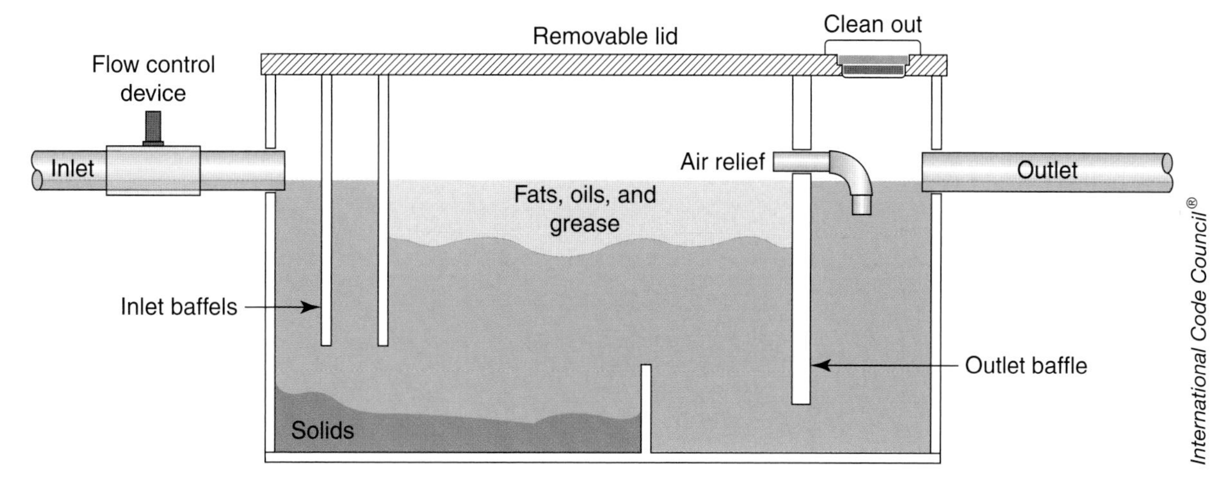

Hydromechanical grease interceptor

1003.3.4 continued

CHANGE SIGNIFICANCE: The plumbing industry has standardized on the terms "hydromechanical" and "gravity" for the two general types of grease interceptors being used. The requirements in Section 1003.3.4 and its subsections have never been intended to apply to gravity-type grease interceptors. The new terminology makes a clear distinction between the two types in order to provide clarity regarding the applicable requirements for each type of grease interceptor.

The PDI G102 standard, "Testing and Certification for Grease Interceptors with Fats, Oil, and Grease (FOG) Sensing and Alarm Devices," expands on the already recognized PDI G101 by including the testing and certification of alarm devices that can be provided on interceptors already complying with PDI G101. The alarm device on a hydromechanical grease interceptor monitors the level of the grease captured in the unit and provides both a loud audible signal and a visible signal when the accumulated (FOG) in the interceptor needs to be removed.

1105

Roof Drain Strainers

CHANGE TYPE: Modification

CHANGE SUMMARY: Outdated code requirements have been replaced with new provisions that properly address the installation and sizing of roof drains.

2012 CODE: ~~**1105.1 Strainers.** Roof drains shall have strainers extending not less than 4 inches (102 mm) above the surface of the roof immediately adjacent to the roof drain. Strainers shall have an available inlet area, above roof level, of not less than one and one-half times the area of the conductor or leader to which the drain is connected.~~

~~**1105.2 Flat decks.** Roof drain strainers for use on sun decks, parking decks and similar areas that are normally serviced and maintained shall comply with Section 1105.1 or shall be of the flat-surface type, installed level with the deck, with an availableinlet area not less than two times the area of the conductor or leader to which the drain is connected.~~

1105.1 General. Roof drains shall be installed in accordance with the manufacturer's instructions. The inside opening for the roof drain shall not be obstructed by the roofing membrane material.

CHANGE SIGNIFICANCE: Prescriptive requirements addressing roof drain strainers have been deleted and replaced with a reference to the manufacturer's instructions. The provisions that were deleted predate the development of the current roof drain standard, ASME A112.21.2M. The original requirements appeared in the ANSI A40.8-1955 standard and were the basis of the legacy codes. With the reference to ASME A112.21.2M in Section 1102.6, the sizing and design of the roof drain is now regulated by the standard. The following table provides a comparison of the sizing of strainers based on Section 1105.1 of the 2009 IPC and ASME A112.21.2M. It is obvious that the 2009 code language does not require nearly the flow area of the strainer as compared to what the current standard requires.

1105 continues

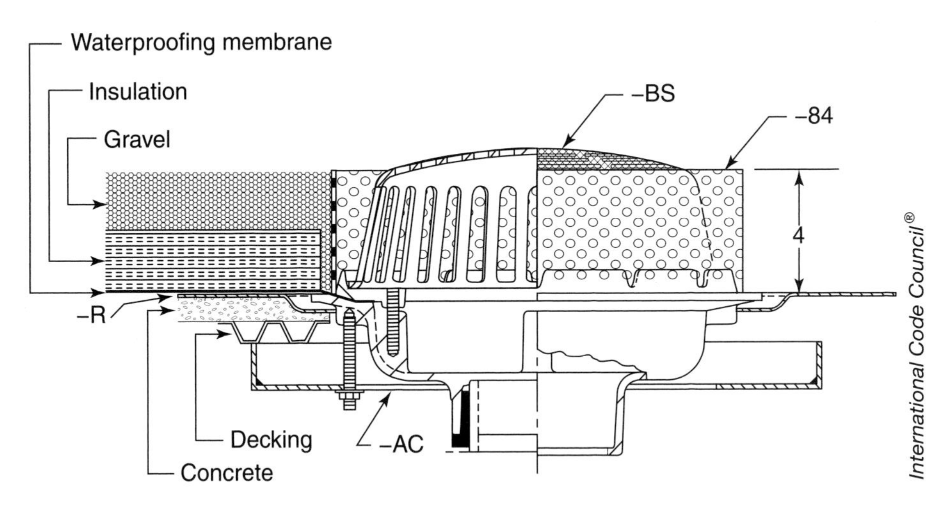

Specification cutaway view of a roof drain installation *(Courtesy of Zurn Industries, LLC)*

1105 continued

TABLE 11-1 Dome Area Comparison Table

Pipe Size	Inside Diameter	Inside Area	Min. Dome Area Per ASME A112.6.4	IPC Minimum Dome Area
(inches)	(inches)	(sq inches)	(sq inches)	(sq inches)
2	1.96	3.02	18	4.53
3	2.96	6.88	25	10.32
4	3.94	12.19	36	18.29
5	4.94	19.17	50	28.75
6	5.94	27.71	70	41.57

By referencing ASME A112.21.2M in Section 1102.6, Section 1105.1 only has to reference the instructions of the manufacturer for regulating the installation. One of the most commonly identified violations in roof drain installations has been the overlapping of the roof membrane into the roof drain. Contractors have been known to cut a smaller opening in the roof membrane than the size of the roof drain. The additional language in Section 1105.1 emphasizes that, after installation, the inside opening of the roof drain outlet must not be blocked by roofing membrane materials.

1107

Siphonic Roof Drainage Systems

CHANGE TYPE: Addition

CHANGE SUMMARY: New requirements have been added to address the design of siphonic roof drainage systems by referencing the standard ASPE 45 for design of the system and ASME A112.6.9 for use of the roof drain.

2012 CODE:

SECTION 1107 SIPHONIC ROOF DRAINAGE SYSTEMS

1107.1 General. Siphonic roof drainage systems shall be designed in accordance with ASME A112.6.9 and ASPE 45.

CHANGE SIGNIFICANCE: The new technology of siphonic roof drainage systems originated in Europe and is being used with much success throughout the world. It has been found that cost benefits over traditional roof drainage methods can be significant depending on the material used, project size, and project complexity. Siphonic drainage is also ideal for rainwater harvesting due to the minimal number of downspouts. Fewer downspouts will make it easier to transport the rainwater from the roof to a retention area, which can include rain barrels, cisterns, drainage ponds, and other storage tanks. The system has the potential to provide many green benefits to the building owner.

Siphonic roof drainage systems are made up of roof drains with air baffles complying with ASME A112.6.9. The air baffle prevents vortex flow while restricting air from entering the vertical tailpipe and horizontal collector piping. As water forces air out, a negative head pressure is created in the collector pipe and the water is sucked off the roof. The high flow capacities and velocities in siphonic systems can be advantageous to the designer, resulting in flexibility in the placement of stacks, smaller pipe diameters providing equivalent flow rates, no pitch requirement of piping to induce flow, and easier coordination of piping with other building elements. Conventional systems operate on a different hydraulic principle than siphonic systems. Siphonic roof drain systems require a higher level of technical understanding by the engineer.

1107 continues

Typical siphonic main roof drain *(Courtesy of Zurn Industries, LLC)*

Typical siphonic overflow roof drain *(Courtesy of Zurn Industries, LLC)*

1107 continued

It is critical that a siphonic roof drainage system is inspected and tested to ensure the drawings and specifications have been complied with as submitted by the designer. The design calculations that have been prepared by the engineer are authentic for specific pipe material, surface roughness, inner diameters, thermal expansion properties, and fittings. The plumbing contractor is not to provide substitutes for the specified pipe material, pipe size, or fittings without approval of the design engineer and completion of revised calculations. The use of design software for siphonic roof drainage systems is highly recommended, as it has become the prevailing standard of care in the plumbing industry. There are other acceptable alternate methods that also can be used. As with any engineered method, the designer shall provide sufficient detail on the drawings to instruct the plumber or installer on the size, orientation, and support of the piping, fittings, and drains.

Design standard ASPE 45 requires siphonic roof drainage systems to be inspected and tested to ensure compliance with the drawings and specifications prepared by the design engineer. These standards also require the designer or designated inspector to periodically inspect and observe the installation of a siphonic roof drainage system. All discrepancies are to be brought to the immediate attention of the plumbing contractor in writing and filed with the code official having jurisdiction. The designer is required to submit a final report in writing to the code official upon completion of the installation. The code official's notice of approval for the installation should not be issued until the designer has issued the written certification. A complete operational test of the siphonic roof drainage system (i.e., a complete flow test) is not practical to implement and is also not a requirement under Section 312 of the *International Plumbing Code* (IPC) for any other drainage system. A complete flow test of a siphonic roof drain system is not required. ASME A112.6.9 and ASPE 45 standards should be consulted for additional information on the design and installation requirements for siphonic roof drainage systems.

CHANGE TYPE: Addition

Chapter 13

Gray-Water Recycling Systems

CHANGE SUMMARY: Provisions addressing gray-water recycling systems have been relocated from Appendix C to a new Chapter 13 in the body of the code.

2012 CODE:

Chapter 13 (Gray-Water Recycling System)

***Reader's Note:** *Gray water has been added to Chapter 2 (Definitions).*

> **Gray Water.** Waste discharged from lavatories, bathtubs, showers, clothes washers, and laundry trays.

***Reader's Note:** *An exception has been added to Section 301.3 to allow gray-water-producing fixtures to be connected to a gray-water collection system where a gray-water recycling system is to be installed.*

301.3 Connections to Drainage System. ~~All~~ Plumbing fixtures, drains, appurtenances and appliances used to receive or discharge liquid wastes or sewage shall be directly connected to the sanitary drainage system of the building or premises, in accordance with the requirements of this code. This section shall not be construed to prevent indirect waste systems required by Chapter 8.

Chapter 13 continues

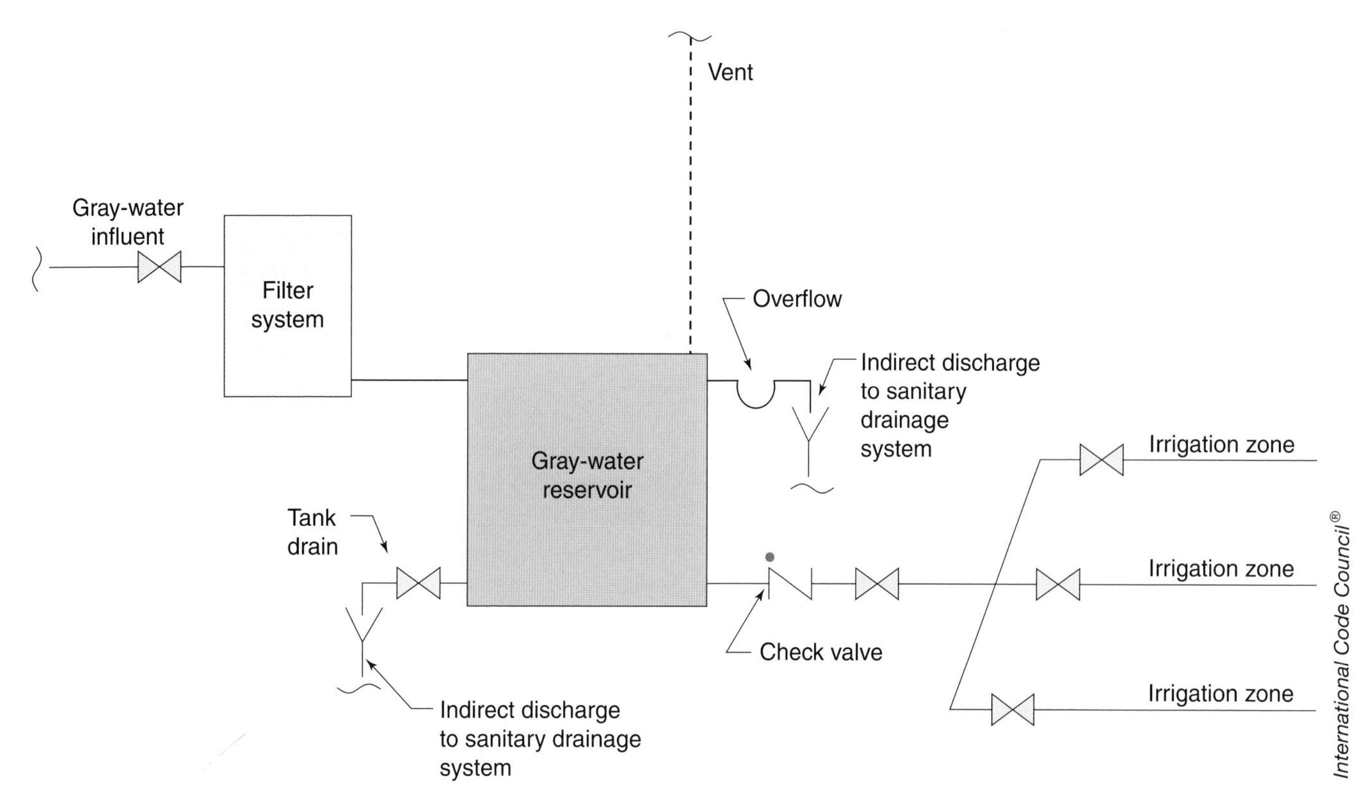

Gray-water recycling system for subsurface landscape irrigation systems

Chapter 13 continued

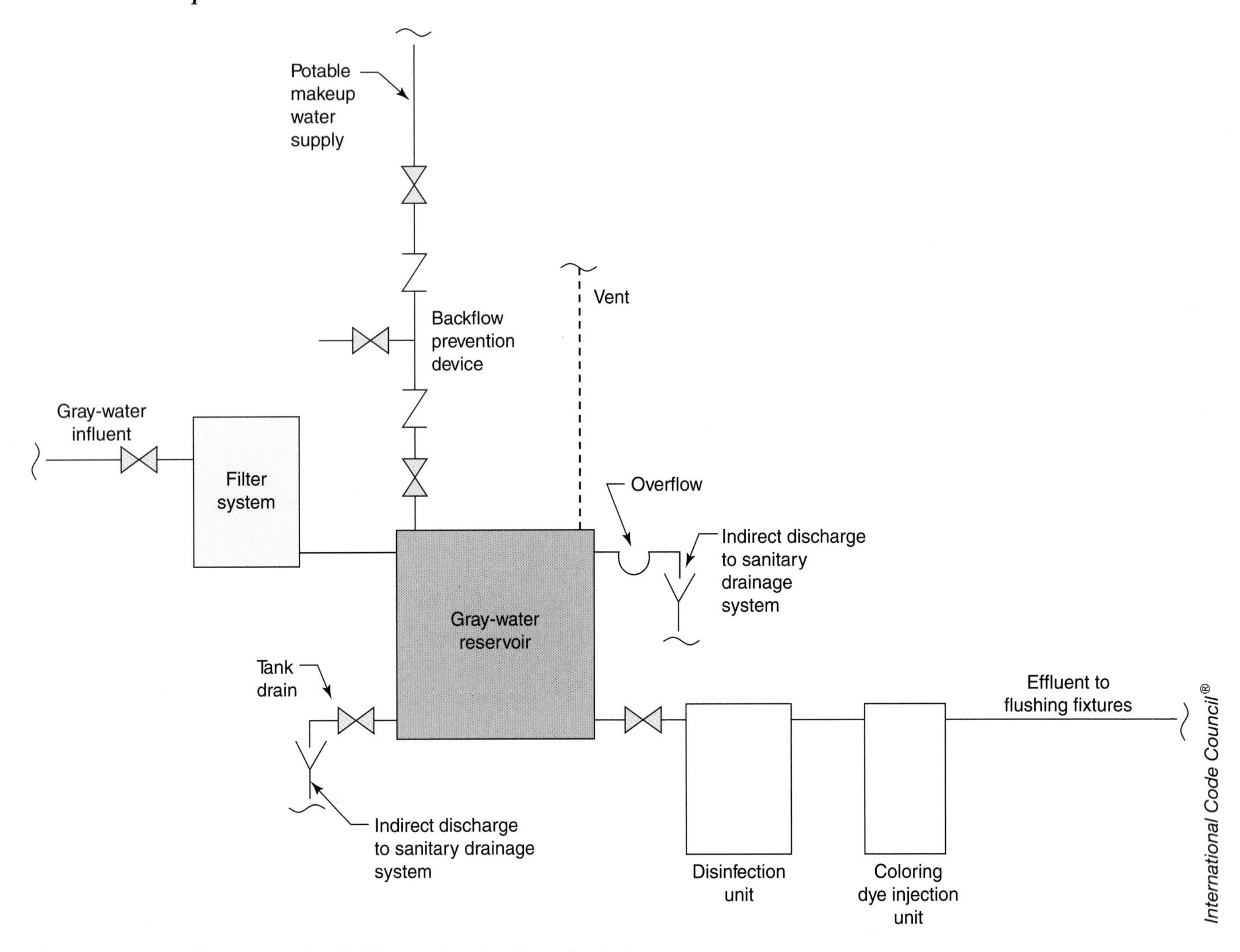

Gray-water recycling system for flushing water closets and urinals

Exceptions: Bathtubs, showers, lavatories, clothes washers, and laundry trays shall not be required to discharge to the sanitary drainage system where such fixtures discharge to an approved gray water system for flushing of water closets and urinals or for subsurface landscape irrigation.

~~APPENDIX C~~ CHAPTER 13
GRAY-WATER RECYCLING SYSTEMS
SECTION ~~C101~~ 1301 GENERAL

~~C101.1~~ 1301.1 Scope. The provisions of ~~this appendix~~ Chapter 13 shall govern the materials, design, construction and installation of gray-water systems for flushing of water closets and urinals and for subsurface landscape irrigation. See Figures 1301.1(1) and 1301.1(2).

***Reader's Note:** *All of the remaining sections of Appendix C have been relocated with their complete text and renumbered accordingly in the new Chapter 13.*

102.3

Maintenance

202

Environmental Air

306.5

Equipment and Appliances on Roofs or Elevated Structures

308.5

Labeled Assemblies

401.4

Intake Opening Location

TABLE 403.3

Minimum Ventilation Rates for Nail Salons

404.1

Enclosed Parking Garages

501.2, 506.4

Independent Exhaust Systems Required

505.1

Domestic Kitchen Exhaust Systems

506.3.7.1

Grease Reservoirs

506.3.8

Grease Duct Cleanouts and Other Openings

506.3.9

Grease Duct Horizontal Cleanouts

506.3.10

Underground Grease Duct Installation

506.3.11.2

Field Applied Grease Duct Enclosure

507.2

Type I or Type II Hood Required

507.2.1

Type I Hoods

507.2.1.1

Operation of Type I Hoods

507.2.1.2

Exhaust Flow Rate Label for Type I Hoods

507.2.2

Type II Hoods

507.10

Hoods Penetrating a Ceiling

510.7

Fire Suppression Required for Hazardous Exhaust Ducts

601.4

Contamination Prevention in Plenums

602.2.1

Materials within Plenums

603.7

Rigid Duct Penetrations

603.9

Duct Joints, Seams and Connections

603.17

Air Dispersion Systems

805.3

Factory Built Chimney Offsets

901.4

Fireplace Accessories

928

Evaporative Cooling Equipment

1101.10

Locking Access Port Caps

1105.6 and 1105.6.3

Machinery Room Ventilation

1106.4

Flammable Refrigerants

102.3

Maintenance

CHANGE TYPE: Modification

CHANGE SUMMARY: ASHRAE/ACCA/ANSI Standard 180 is now specified for the inspection for maintenance of an HVAC system.

2012 CODE: 102.3 Maintenance. Mechanical systems, both existing and new, and parts thereof shall be maintained in proper condition in accordance with the original design and in a safe and sanitary condition. Devices or safeguards which are required by this code shall be maintained in compliance with the code edition under which they were installed. The owner or owner's designated agent shall be responsible for maintenance of mechanical systems. To determine compliance with this provision, the code official shall have the authority to require a mechanical system to be re-inspected. The inspection for maintenance of HVAC systems shall be done in accordance with ASHRAE/ACCA/ANSI Standard 180.

CHANGE SIGNIFICANCE: Through the addition of the requirement that the inspection for maintenance of HVAC systems be done in accordance with ASHRAE/ACCA/ANSI Standard 180, the code official is provided with direct guidance as to what parts of the HVAC system will have to be maintained. The referenced standard has specific requirements for maintaining duct systems, air handlers, boilers, chillers, water towers, and numerous other components of an HVAC system. The standard also specifies at what intervals these components are to be inspected.

ANSI/ASHRAE/ACCA Standard 180-2008

ASHRAE/ACCA STANDARD

Standard Practice for Inspection and Maintenance of Commercial HVAC Systems

Approved by the ASHRAE Standards Committee on ______, 2008; by the ASHRAE Board of Directors on ______, 2008; by Air Conditioning Contractors of America on ______, 2008; and by the American National Standards Institute on ______, 2008.

ASHRAE Standards are scheduled to be updated on a five-year cycle; the date following the standard number is the year of ASHRAE Board of Directors approval.

The latest copies may be purchased from ASHRAE or ACCA. From ASHRAE at www.ashrae.org, by emailing orders@ashrae.org, by faxing 404-321-5478, or by calling Customer Service at 404-636-8400 (worldwide) or toll free at 1-800-527-4723 (for orders in US and Canada). From ACCA at www.acca.org, by faxing 703-575-4449, or by calling 703-575-4477.

© Copyright 2007 ASHRAE, Inc., and ACCA.

ISSN 1041-2336

Approved American National Standard

ANSI

www.ansi.org

American Society of Heating, Refrigerating and Air-Conditioning Engineers, Inc.
1791 Tullie Circle NE, Atlanta, GA 30329 www.ashrae.org

ACCA

Air Conditioning Contractors of America
2800 Shirlington Rd., Ste 300, Arlington, VA 22206 www.acca.org

ASHRAE/ACCA standards *(Courtesy of ACCA)*

202 Environmental Air

CHANGE TYPE: Clarification

CHANGE SUMMARY: The definition of *environmental air* has been expanded through the addition of parking garage exhaust.

2012 CODE: 202 Environmental Air: Air that is conveyed to or from occupied areas through ducts which are not part of the heating or air-conditioning system, such as ventilation for human usage, domestic kitchen range exhaust, bathroom exhaust, ~~and~~ domestic clothes dryer exhaust and parking garage exhaust.

CHANGE SIGNIFICANCE: The addition of parking garage exhaust to the environmental air definition is intended to clarify any inconsistencies in enforcement. Previously, it was unclear if parking garage exhaust was to be considered product-conveying or environmental air. The clarification will be helpful in the determination of where a parking garage exhaust system must terminate based on the requirements of Section 501.2.1 in the IMC.

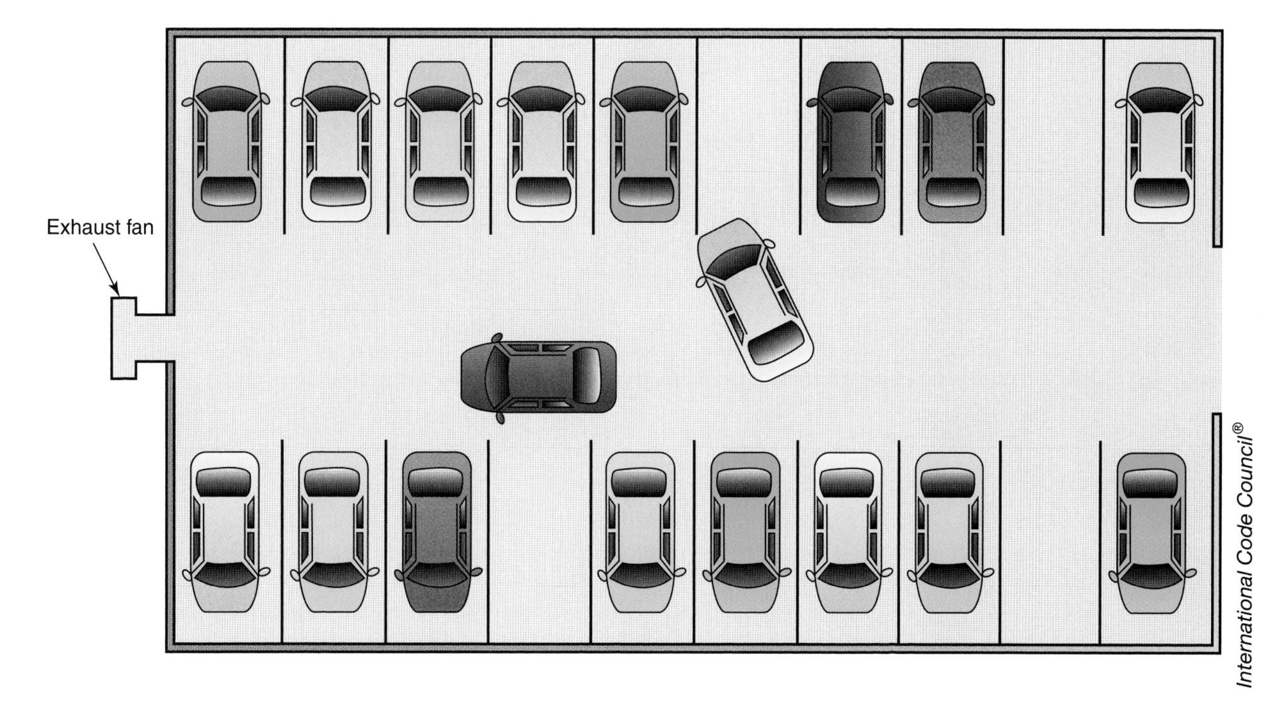

Parking garage

306.5

Equipment and Appliances on Roofs or Elevated Structures

CHANGE TYPE: Modification

CHANGE SUMMARY: It has been clarified that permanent access is required to equipment and appliances on a roof or elevated structure higher than 16 feet above grade, and required clearances are now provided to assure access to ladders required for access to roofs or elevated structures.

2012 CODE: 306.5 Equipment and Appliances on Roofs or Elevated Structures. Where equipment requiring access ~~and~~ or appliances are ~~installed on roofs or elevated structures at a height exceeding 16 feet (4877 mm), such access shall be provided by a permanent approved means of access, the extent of which shall be from grade or floor level to the equipment and appliances' level service space~~ located on an elevated structure or the roof of a building such that personnel will have to climb higher than 16 feet (4877 mm) above grade to access such equipment or appliances, an interior or exterior means of access shall be provided. Such access shall not require climbing over obstructions greater than 30 inches (762 mm) in height or walking on roofs having a slope greater than 4 units vertical in 12 units horizontal (33-percent slope). Such access shall not require the use of portable ladders. Where access involves climbing over parapet walls, the height shall be measured to the top of the parapet wall.

Permanent ladders installed to provide the required access shall comply with the following minimum design criteria:

1. The side railing shall extend above the parapet or roof edge not less than 30 inches (762 mm).
2. Ladders shall have rung spacing not to exceed 14 inches (356 mm) on center. The upper-most rung shall be a maximum of

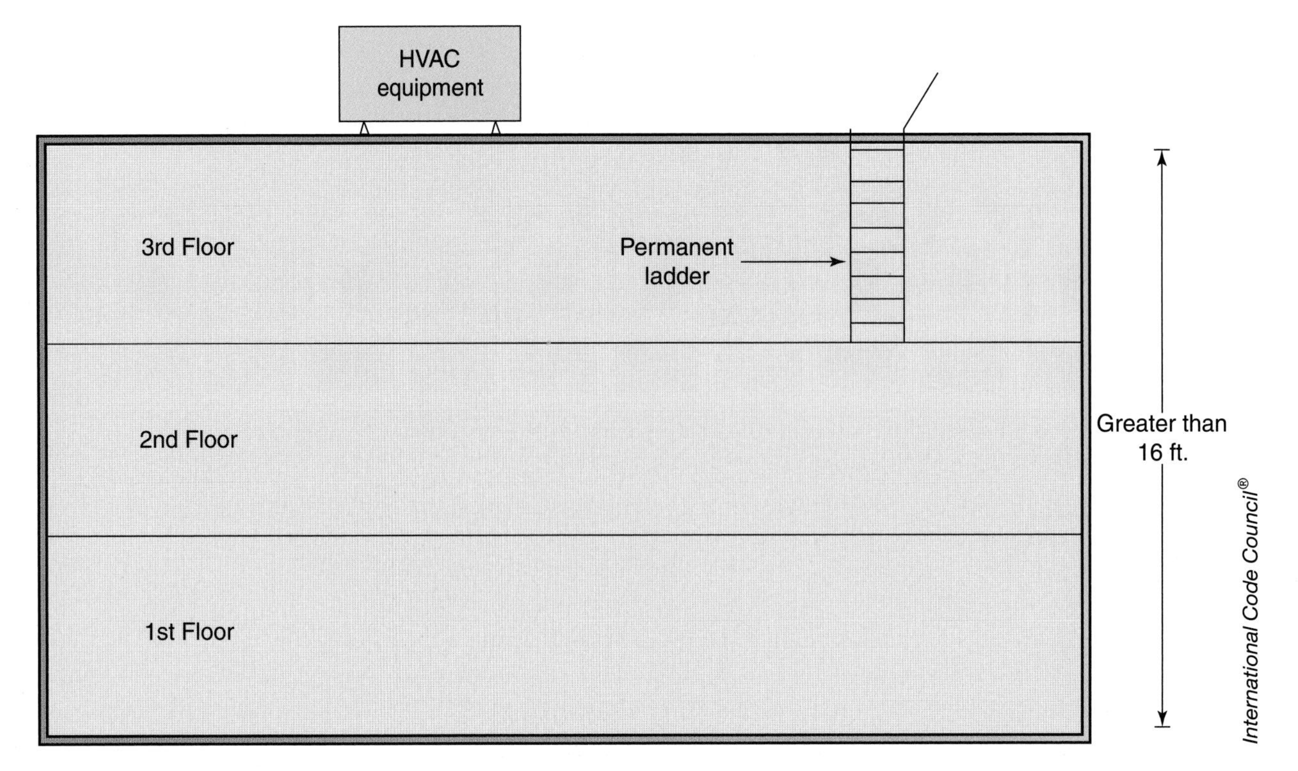

Roof access

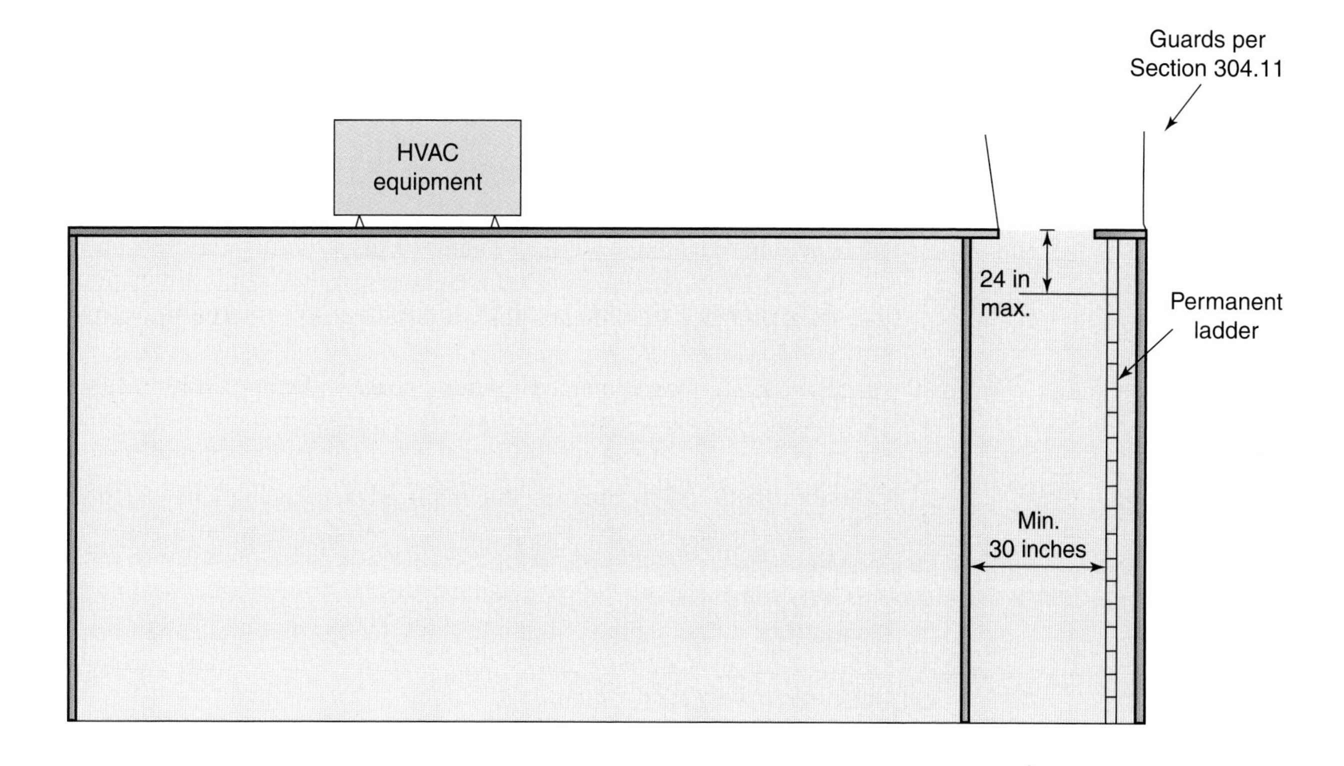

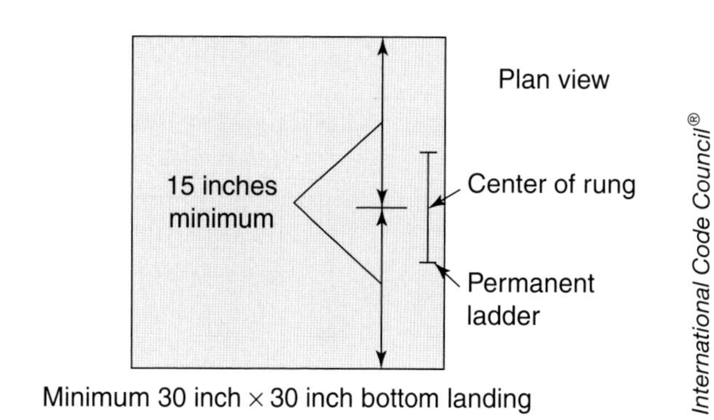

International Code Council®

Access ladder

24 inches (610 mm) below the upper edge of the roof hatch, roof or parapet, as applicable.

3. Ladders shall have a toe spacing not less than 6 inches (152 mm) deep.
4. There shall be a minimum of 18 inches (457 mm) between rails.
5. Rungs shall have a minimum 0.75-inch (19 mm) diameter and be capable of withstanding a 300-pound (136.1kg) load.
6. Ladders over 30 feet (9144 mm) in height shall be provided with offset sections and landings capable of withstanding 100 pounds per square foot (488.2 kg/m2). Landing dimensions shall be not less than 18 inches (457 mm) and not less than the width of the ladder served. A guard rail shall be provided on all open sides of the landing.
7. Climbing clearance. The distance from the centerline of the rungs to the nearest permanent object on the climbing side of the

306.5 continues

306.5 continued

ladder shall be a minimum of 30 inches (762 mm) measured perpendicular to the rungs. This distance shall be maintained from the point of ladder access to the bottom of the roof hatch. A minimum clear width of 15-inches (381 mm) shall be provided on both sides of the ladder measured from the midpoint of and parallel with the rungs except where cages or wells are installed.

8. Landing required. The ladder shall be provided with a clear and unobstructed bottom landing area having a minimum dimension of 30 inches (762 mm) by 30 inches (762 mm) centered in front of the ladder.
~~7~~9. Ladders shall be protected against corrosion by approved means.
10. Access to ladders shall be provided at all times.

Catwalks installed to provide the required access shall be not less than 24 inches (610 mm) wide and shall have railings as required for service platforms.

Exception: This section shall not apply to Group R-3 occupancies.

CHANGE SIGNIFICANCE: Where a piece of equipment or an appliance that requires access is located on a roof or elevated structure more than 16 feet above grade level, it has been clarified that a means of access must be provided. If an appliance is located on the roof of a multistory building and the roof access is through a roof hatch opening on the top story of the building, a permanent ladder is required from the floor level to the top roof hatch. The same requirements would apply if a piece of equipment or an appliance were located on an elevated platform on the roof of a building. A permanent ladder would be required to the appliance or equipment located on the platform. The modifications to the permanent ladder criteria will eliminate several hazards that have commonly occurred with roof or elevated-structure access ladders. Previous code editions did not specify how far away a ladder could terminate from a roof access opening or a minimum required distance from the front or sides of a ladder to an obstruction. A minimum landing area is also now required at the bottom of a ladder, requiring ladders to be accessible at all times.

308.5

Labeled Assemblies

CHANGE TYPE: Modification

CHANGE SUMMARY: Allowable clearance reductions must now be based on listed and labeled reduced-clearance protective assemblies in accordance with UL 1618.

2012 CODE: **308.5 Labeled Assemblies:** The allowable clearance reduction shall be based on an approved reduced clearance protective assembly that ~~has been tested and bears the label of an approved agency~~ is listed and labeled in accordance with UL 1618.

CHANGE SIGNIFICANCE: The previous edition of the IMC stated that clearance protective assemblies had to be tested and have a label from an approved agency. A referenced standard is now specified as the basis for the required label and listing. UL Standard 1618, *Wall Protectors, Floor Protectors and Hearth Extensions*, has construction and performance requirements to evaluate wall and floor protective assemblies, as well as hearth extensions that are used with heat-producing appliances.

Listed and labeled clearance protection assembly *(Courtesy of Northwest Stoves)*

401.4

Intake Opening Location

CHANGE TYPE: Modification

CHANGE SUMMARY: The minimum clearance between an air intake opening and any public way is now measured from the opening to the lot line, not to the centerline of the public way.

2012 CODE: 401.4 Intake Opening Location. Air intake openings shall comply with all of the following:

1. Intake openings shall be located a minimum of 10 feet (3048 mm) from lot lines or buildings on the same lot. ~~Where openings front on a street or public way, the distance shall be measured to the centerline of the street or public way.~~
2. Mechanical and gravity outdoor air intake openings shall be located not less than 10 feet (3048 mm) horizontally from any hazardous or noxious contaminant source, such as vents, streets, alleys, parking lots and loading docks, except as specified in Item 3 or Section 501.2.1. Outdoor air intake openings shall be permitted to be located less than 10 feet (3048 mm) horizontally from streets, alleys, parking lots and loading docks provided that the openings are located not less than 25 feet (7620 mm) vertically above such locations. Where openings front on a street or public way the distance shall be measured from the closest edge of the street or public way.
3. Intake openings shall be located not less than 3 feet (914 mm) below contaminant sources where such sources are located within 10 feet (3048 mm) of the opening.
4. Intake openings on structures in flood hazard areas shall be at or above the ~~design flood level~~ elevation required by Section 1612 of the *International Building Code* for utilities and attendant equipment.

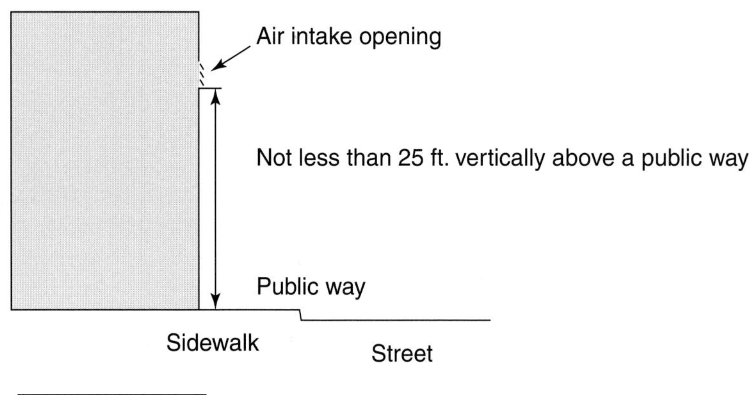

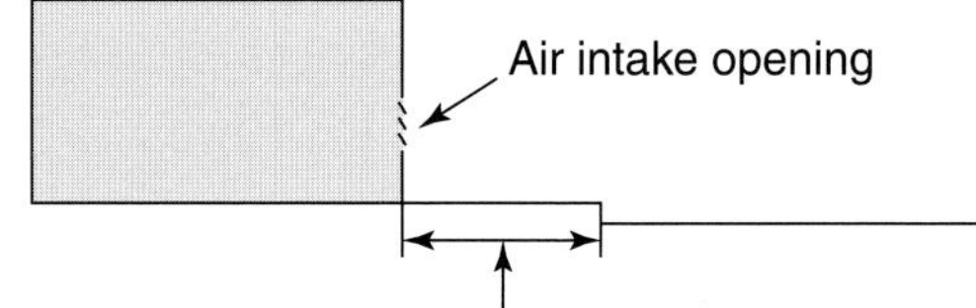

Where openings front on a street or public way, the distance shall be measured from the closest edge of the street or public way

Air intake location

International Code Council®

CHANGE SIGNIFICANCE: Air intake openings for ventilation purposes must be located an established distance from any adjoining street or public way. Previously, the measurement was to be taken to the centerline of the street or public way. In order to maintain adequate separation, the measurement is now taken to all lot lines, including those that front on a street or public way. In a related change, mechanical and gravity air intake openings may now be located closer than 10 feet from streets, alleys, parking lots, and loading docks if the opening is located at least 25 feet vertically above these locations.

Table 403.3

Minimum Ventilation Rates for Nail Salons

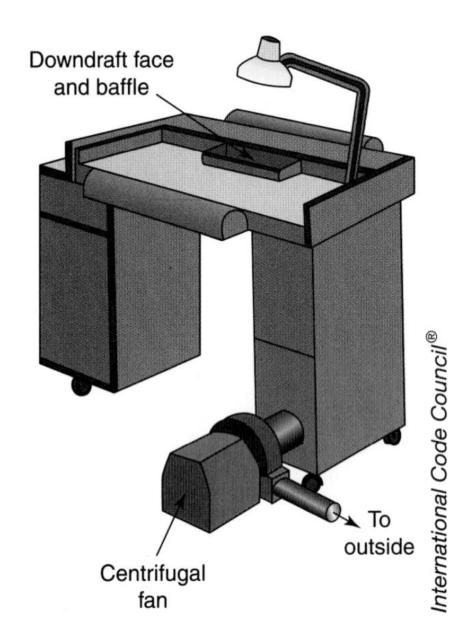

Nail salon station

CHANGE TYPE: Modification

CHANGE SUMMARY: Nail stations in nail salons must now each be provided with a source capture system.

2012 CODE:

TABLE 403.3 Minimum Ventilation Rates

Occupancy Classification	~~People Outdoor Airflow Rate In Breathing Zone R_pCFM/Person~~ Occupant Density #/1000 ft [2a]	~~Area Outdoor Airflow Rate In Breathing Zone R_aCFM/FT[2a]~~ People Outdoor Airflow Rate In Breathing Zone R_pCFM/Person	~~Default Occupant Density #/1000 ft[2a]~~ Area Outdoor Airflow Rate In Breathing Zone R_a CFM/FT[2a]	Exhaust Airflow Rate CFM/FT[2a]
Beauty ~~and nail~~ salons [b,~~h~~]	20	0.12	25	0.6
Nail Salons[b,h]	25	20	0.12	0.6

e. Rates are per water closet or urinal. ~~The higher rate shall be provided where periods of heavy use are expected to occur, such as, toilets in theaters, schools, and sports facilities. The lower rate shall be permitted where periods of heavy use are not expected.~~ The higher rate shall be provided where the exhaust system is designed to operate intermittently. The lower rate shall be permitted only where the exhaust system is designed to operate continuously while occupied.

f. Rates are per room unless otherwise indicated. The higher rate shall be provided where the exhaust system is designed to operate intermittently. The lower rate shall be permitted only where the exhaust system is designed to operate continuously while occupied ~~normal hours of use~~.

h. For nail salons~~, the required exhaust shall include ventilation tables or other systems that capture the contaminants and odors at their source and are capable of exhausting a minimum of 50 cfm per station~~ each nail station shall be provided with a source capture system capable of exhausting not less than 50 cfm per station.

[No changes to portions of table and footnotes not shown.]

CHANGE SIGNIFICANCE: Footnote "h" to Table 403.3 has been modified to require nail salons to have a source capture system at each nail station. Based on the definition of a source capture system, the exhaust from a station in a nail salon is required to capture the air contaminates at their source and terminate them to the outdoor atmosphere. A minimum exhaust rate of 50 cfm is required at each station. Other clarifications were made, including changes to the headings in the table as well as footnotes "d" and "f." The footnotes have been clarified by addressing when the higher and lower exhaust rates specified in the table for toilet rooms in public spaces are to be used.

404.1

Enclosed Parking Garages

CHANGE TYPE: Modification

CHANGE SUMMARY: The mechanical ventilation systems required in enclosed parking garages are now permitted to be operated automatically by carbon monoxide detectors.

2012 CODE: 404.1 Enclosed Parking Garages. Mechanical ventilation systems for enclosed parking garages shall be permitted to operate intermittently in accordance with Item 1, Item 2 or both. ~~where~~

1. The system ~~is~~ shall be arranged to operate automatically upon detection of vehicle operation or the presence of occupants by approved automatic detection devices.
2. The system shall be arranged to operate automatically by means of carbon monoxide detectors applied in conjunction with nitrogen dioxide detectors. Such detectors shall be installed in accordance with their manufacturers' recommendations.

CHANGE SIGNIFICANCE: Historically, enclosed parking garage mechanical exhaust systems have been required to operate continuously or, as an alternative, to operate intermittently upon the detection of vehicle operation or the presence of occupants. An additional option for operation of the ventilation system is now available. Carbon monoxide detectors applied in conjunction with nitrogen dioxide detectors may now also be utilized to automatically operate parking garage exhaust systems, provided they are installed in accordance with the recommendations of the manufacturer.

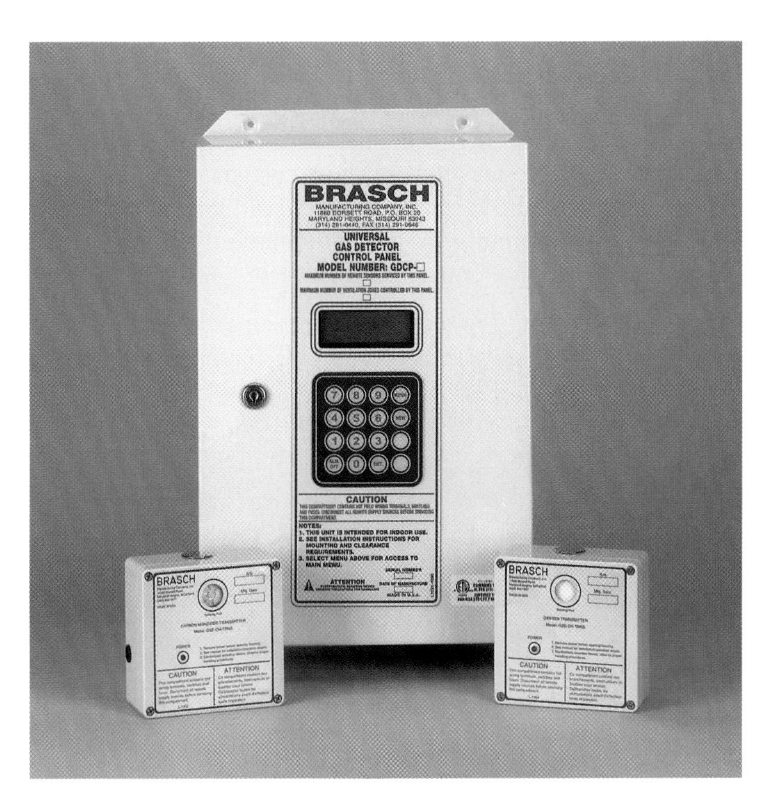

Carbon monoxide detector *(Coutesy of Brasch Manufacturing Co.)*

501.2, 506.4

Independent Exhaust Systems Required

CHANGE TYPE: Modification

CHANGE SUMMARY: Those locations where an independent exhaust system is required are now established in a single code provision.

2012 CODE: **501.2 Independent System Required.** Single or combined mechanical exhaust systems for environmental air shall be independent of all other exhaust systems. Dryer exhaust shall be independent of all other systems. Type I exhaust systems shall be independent of all other exhaust systems except as provided in Section 506.3.5. Single or combined Type II exhaust systems for food-processing operations shall be independent of all other exhaust systems. Kitchen exhaust systems shall be constructed in accordance with Section 505 for domestic equipment and Sections 506 through 509 for commercial equipment.

506.4 Ducts Serving Type II Hoods. ~~Single or combined Type II exhaust systems for food-processing operations shall be independent of all other exhaust systems~~. Commercial kitchen exhaust systems serving Type II hoods shall comply with Sections 506.4.1 and 506.4.2.

CHANGE SIGNIFICANCE: Requirements addressing individual exhaust systems have been relocated from other sections in the IMC and placed in a single location. Most exhaust systems are prohibited from being combined with each other. As an example, a bathroom or toilet room exhaust system cannot be combined with a Type II exhaust system for food processing operations. It has been emphasized that clothes dryer exhaust systems and Type I and Type II commercial cooking exhaust systems are to be independent of all other exhaust systems. It should be noted that due to a new change to Section 505.1, domestic range hoods are now required to have independent exhaust systems. Another related change that will have an effect occurs in Section 202, where it now defines parking garage exhaust as environmental air. Therefore, it is possible to combine a bathroom or toilet room exhaust system with a parking garage exhaust system because both systems are considered environmental air. In addition, the requirements for an independent Type II commercial cooking exhaust system previously addressed in Section 506.4 have been relocated to Section 501.2.

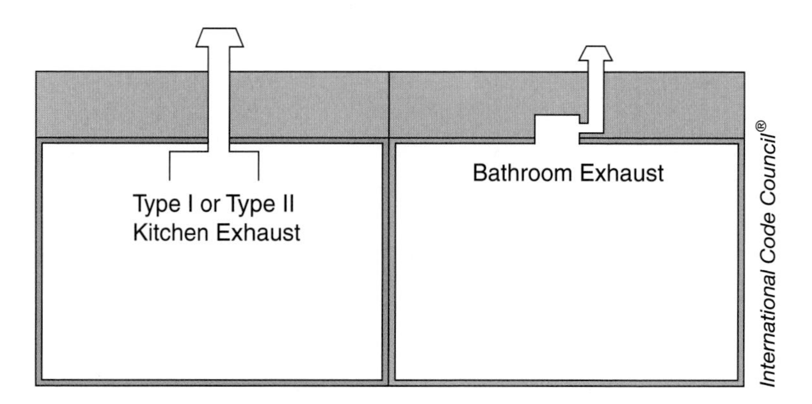

Independent exhaust system

505.1

Domestic Kitchen Exhaust Systems

CHANGE TYPE: Modification

CHANGE SUMMARY: Domestic kitchen exhaust ducts are now required to be independent of all other exhaust systems.

2012 CODE: **505.1 Domestic Systems.** Where domestic range hoods and domestic appliances equipped with downdraft exhaust are located within dwelling units, such hoods and appliances shall discharge to the outdoors through sheet metal ducts constructed of galvanized steel, stainless steel, aluminum or copper. Such ducts shall have smooth inner walls, ~~and~~ shall be air tight, and shall be equipped with a back-draft damper, and shall be independent of all other exhaust systems.

Exceptions:

1. Where installed in accordance with the manufacturer's installation instructions and where mechanical or natural ventilation is otherwise provided in accordance with Chapter 4, listed and labeled ductless range hoods shall not be required to discharge to the outdoors.
2. Ducts for domestic kitchen cooking appliances equipped with downdraft exhaust systems shall be permitted to be constructed of Schedule 40 PVC pipe and fittings provided that the installation complies with all of the following:
   - 2.1. The duct shall be installed under a concrete slab poured on grade.
   - 2.2. The under floor trench in which the duct is installed shall be completely backfilled with sand or gravel.
   - 2.3. The PVC duct shall extend not more than 1 inch (25 mm) above the indoor concrete floor surface.
   - 2.4. The PVC duct shall extend not more than 1 inch.
   - 2.5. The PVC ducts shall be solvent cemented.

CHANGE SIGNIFICANCE: In the past, it has been a common practice to combine bathroom or toilet room exhaust systems with exhaust from a domestic range hood. Manufacturers of these products have not promoted this practice due to the difference in air movement in the different systems. A new provision now requires domestic kitchen exhaust systems to be independent of all other exhaust systems.

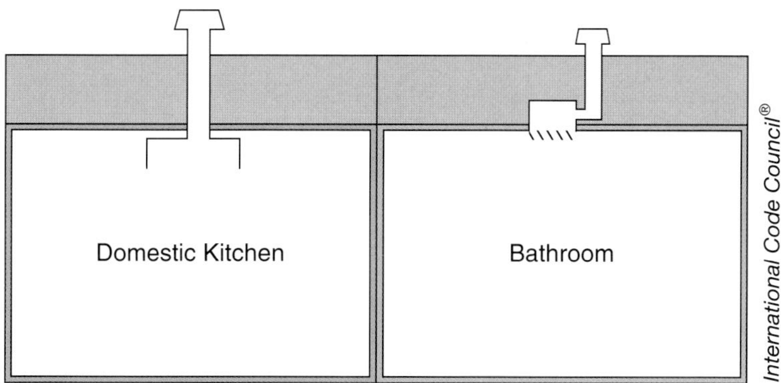

Independent domestic exhaust system

506.3.7.1

Grease Reservoirs

CHANGE TYPE: Addition

CHANGE SUMMARY: Criteria are now provided for the construction of a grease reservoir in a grease duct system where the reservoir is not a manufactured product.

2012 CODE: **506.3.7.1 Grease Reservoirs.** Grease reservoirs shall:

1. Be constructed as required for the grease duct it serves.
2. Be located on the bottom of the horizontal duct or the bottommost section of the duct riser.
3. Have a length and width of not less than 12 inches. Where the grease duct is less than 12 inches in a dimension, the reservoir shall be not more than 2 inches smaller than the duct in that dimension.
4. Have a depth of not less than 1 inch.
5. Have a bottom that is sloped to a point for drainage.
6. Be provided with a cleanout opening constructed in accordance with Section 506.3.8 and installed to provide direct access to the reservoir. The cleanout opening shall be located on a side or on top of the duct so as to permit cleaning of the reservoir.
7. Be installed in accordance with the manufacturer's installation instructions where manufactured devices are utilized.

CHANGE SIGNIFICANCE: Previous editions of the IMC have required a grease duct to be sloped toward an approved grease reservoir, but there have never been any provisions to address how a grease reservoir should be constructed. Prescriptive requirements on how to construct a grease reservoir are now established in Section 506.3.7.1. The seven items listed provide a commonsense type of approach to the construction of a grease reservoir. The provisions not only specify how a grease reservoir is to be constructed, but also allow for the installation of factory-built grease reservoirs.

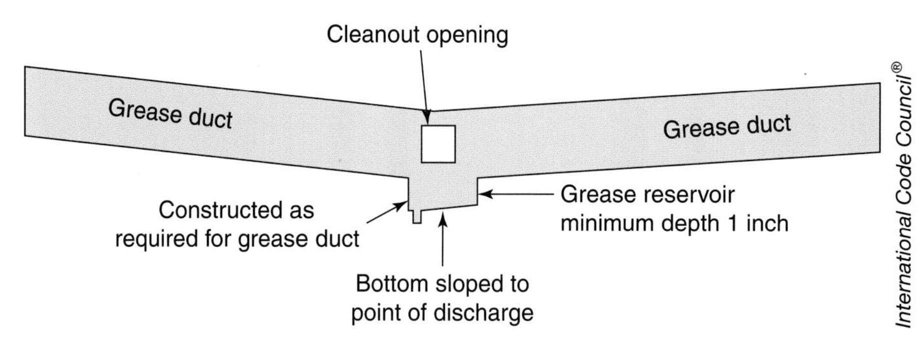

Grease reservoir

506.3.8

Grease Duct Cleanouts and Other Openings

CHANGE TYPE: Modification

CHANGE SUMMARY: In addition to the reformatting of previous criteria for grease duct cleanouts, gasket and sealing materials on grease duct cleanout doors must now be rated at a minimum of 1500°F.

2012 CODE: 506.3.8 Grease Duct Cleanouts and Other Openings. Grease duct cleanouts and openings shall comply with all of the following:

1. Grease ducts shall not have openings except where required for the operation and maintenance of the system.
2. Sections of grease ducts that are inaccessible from the hood or discharge openings shall be provided with cleanout openings.
3. Cleanouts and openings shall be equipped with tight fitting doors constructed of steel having a thickness not less than that required for the duct.
4. Cleanout doors shall be installed liquid tight.
5. Door assemblies including any frames and gaskets shall be approved for the application and shall not have fasteners that penetrate the duct.
6. Gasket and sealing materials shall be rated for not less than 1500 degrees F (815.6 C).
7. Listed door assemblies shall be installed in accordance with the manufacturer's installation instructions.

(2009 code text not shown for clarity)

506.3.8 continues

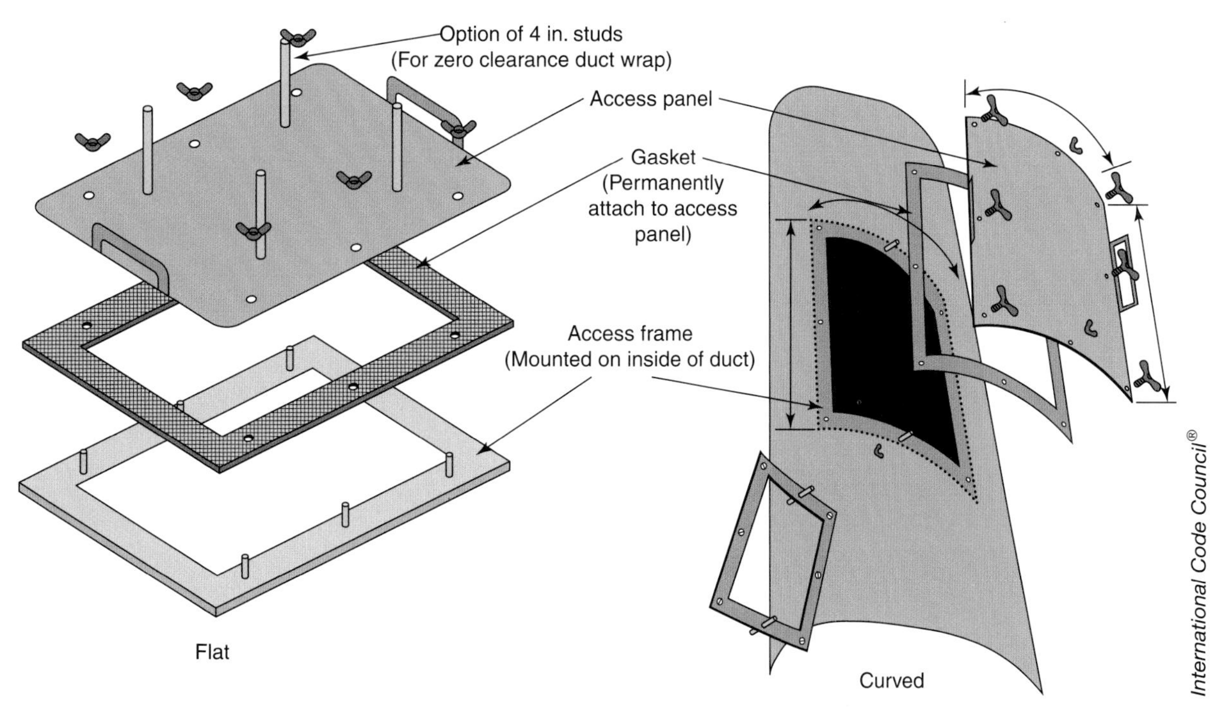

Grease duct cleanout opening

506.3.8 continued

CHANGE SIGNIFICANCE: In order to provide greater ease in understanding the criteria for cleanouts and other openings in grease duct systems, item 6 has been added and specifies a minimum temperature rating for gasket and sealing materials on grease duct cleanout doors. The temperature threshold of at least 1500°F is consistent with the temperature requirements for the gasketing used for the duct-to-hood joint as addressed in Section 506.3.2.2, Exception 1.

506.3.9

Grease Duct Horizontal Cleanouts

CHANGE TYPE: Modification

CHANGE SUMMARY: Criteria for cleanouts serving horizontal grease ducts have been rearranged for ease of use and clarification, and several technical provisions have been added or modified.

2012 CODE: 506.3.9 Grease Duct Horizontal Cleanouts. Cleanouts serving horizontal sections of grease duct shall:

1. Be spaced not more than 20 feet apart.
2. Be located not more than 10 feet from changes in direction that are greater than 45 degrees.
3. Be located on the bottom only where no other locations are available and shall be provided with internal damming of the opening such that grease will flow past the opening without pooling. Bottom cleanouts and openings shall be approved for the application and installed liquid tight.
4. Not be closer than 1 inch from the edges of the duct.
5. Have opening dimensions of not less than 12 inches by 12 inches. Where such dimensions preclude installation, the opening shall be not less than 12 inches on one side and shall be large enough to provide access for cleaning and maintenance.
6. Shall be located at grease reservoirs.

(2009 code text not shown for clarity)

CHANGE SIGNIFICANCE: Provisions addressing cleanouts that serve horizontal grease ducts have been reformatted in a manner that provides for clarity and ease of use. In addition, item 2 now requires a cleanout where a change in direction occurs that is greater than 45 degrees. This cleanout must be located no more than 10 feet from the point the duct changes direction. The minimum distance a cleanout opening must be located from the bottom of a horizontal grease duct has decreased from 1½ inches to 1 inch. The provisions now also include a new requirement mandating a cleanout at grease reservoirs.

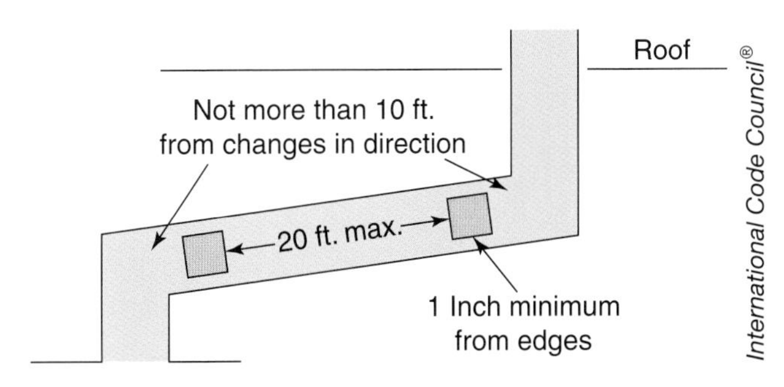

Horizontal grease duct cleanout opening

506.3.10

Underground Grease Duct Installations

CHANGE TYPE: Addition

CHANGE SUMMARY: Grease ducts installed in underground locations are now regulated based upon a number of new provisions.

2012 CODE: **506.3.10 Underground Grease Duct Installation.** Underground grease duct installations shall comply with all of the following:

1. Underground grease ducts shall be constructed of steel having a minimum thickness of 0.0575 inch (1.463mm) (No. 16 gage) and shall be coated to provide protection from corrosion or shall be constructed of stainless steel having a minimum thickness of 0.0450 inch (1.1400 mm) (No.18 gage).
2. The underground duct system shall be tested and approved in accordance with Section 506.3.2.5 prior to coating or placement in the ground.
3. The underground duct system shall be completely encased in concrete with a minimum thickness of 4 inches.
4. Ducts shall slope toward grease reservoirs.
5. A grease reservoir with a cleanout to allow cleaning of the reservoir shall be provided at the base of each vertical duct riser.
6. Cleanouts shall be provided with access to permit cleaning and inspection of the duct in accordance with Section 506.3.
7. Cleanouts in horizontal ducts shall be installed on the topside of the duct.
8. Cleanout locations shall be legibly identified at the point of access from the interior space

CHANGE SIGNIFICANCE: Because of the increase in the number of underground grease duct installations, new provisions have been established to address such installations. Underground grease duct systems are used in food establishments that cook meals at each individual table

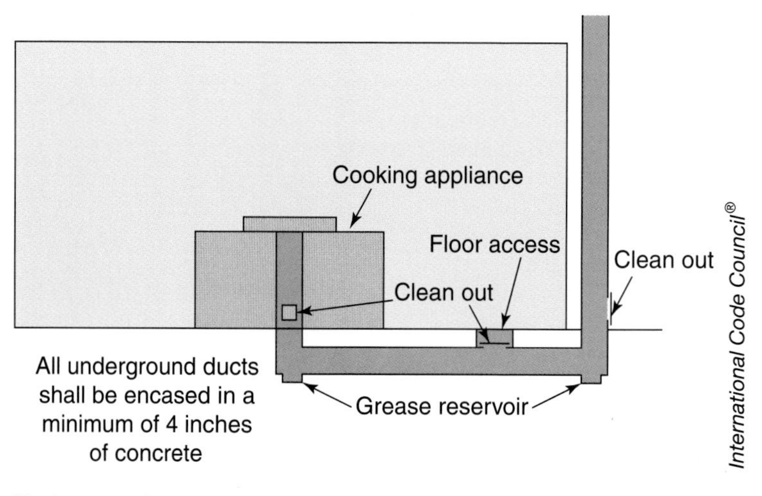

Underground grease duct

where the patrons are sitting. With the increasing popularity of table-top cooking conducted within the dining area, more and more restaurants are installing table-top cooking appliances with integral downdraft exhaust systems. Grease ducts connect to the outlet of the downdraft exhaust system and are routed through the floor and then turn up when they are outside the dining area and go up through the roof. The intent of the new provisions is to provide prescriptive requirements on the installation of the underground grease ducts used for table-top cooking. Although some of the requirements are consistent with those applied to the typical grease duct systems, new provisions require the encasement of underground grease ducts in at least 4 inches of concrete and mandate that reservoirs be provided at the locations of cleanouts.

506.3.11.2

Field-Applied Grease Duct Enclosures

CHANGE TYPE: Clarification

CHANGE SUMMARY: Field-applied grease duct enclosure systems are now specifically prohibited from being used to reduce clearance to combustibles.

2012 CODE: **506.3.11.2 Field-Applied Grease Duct Enclosure.** Commercial kitchen grease ducts constructed in accordance with Section 506.3.1 shall be enclosed by field-applied grease duct enclosure that is a listed and labeled material, system, product, or method of construction specifically evaluated for such purpose in accordance with ASTM E2336. The surface of the duct shall be continuously covered on all sides from the point at which the duct originates to the outlet terminal. Duct penetrations shall be protected with a through-penetration fire-stop system classified in accordance with ASTM E814 or UL 1497 and having a "F" and "T" rating equal to the fire-resistance rating of the assembly being penetrated. Such systems shall be installed in accordance with the listing and the manufacturer's installation instructions. Partial application of a field-applied grease duct enclosure system shall not be installed for the sole purpose of reducing clearance to combustibles at isolated sections of grease duct. Exposed duct-wrap systems shall be protected where subject to physical damage.

CHANGE SIGNIFICANCE: The basis of this code change comes from the misuse of a listed product. Field-applied grease duct enclosure systems serving Type I hoods have not been tested as partial systems for the purpose of reducing the clearance of a grease duct to combustible material. The misapplication of this duct enclosure system usually occurs where a grease duct does not require an enclosure per IMC Section 506.3.11.4 and the grease duct penetrates a roof assembly that is constructed of combustible material or there is combustible material on the decking. In order to reduce the clearance between the grease duct and the combustible roof assembly or combustible material on the decking, the field-applied grease duct enclosure system is often installed around the grease duct at the roof penetration. However, this method of clearance reduction is unacceptable, as the duct enclosure system was never intended to be used this way, nor was it tested for this type of use.

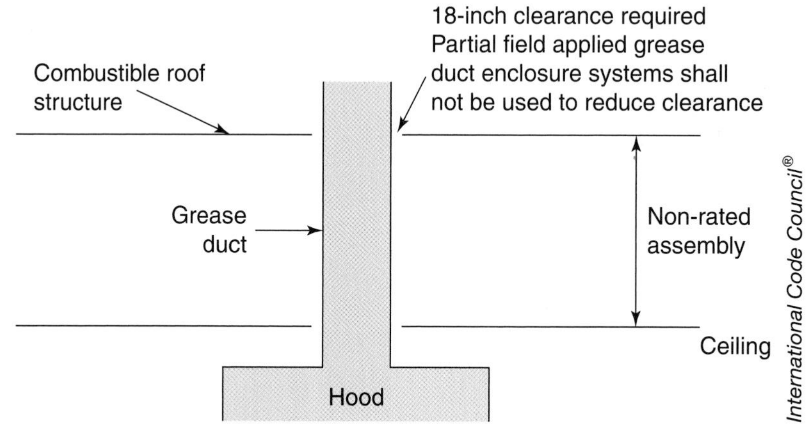

Grease duct clearance

507.2

Type I or Type II Hood Required

CHANGE TYPE: Modification

CHANGE SUMMARY: A Type I or Type II commercial kitchen hood is not required for appliances with listed integral downdraft exhaust systems.

2012 CODE: 507.2 Where Required. A Type I or Type II hood shall be installed at or above all commercial cooking appliances in accordance with Sections 507.2.1 and 507.2.2. Where any cooking appliance under a single hood requires a Type I hood, a Type I hood shall be installed. Where a Type II hood is required, a Type I or Type II hood shall be installed.

> **Exception:** Where cooking appliances are equipped with integral down-draft exhaust systems and such appliances and exhaust systems are listed and labeled for the application, in accordance with NFPA 96, a hood shall not be required at or above them.

CHANGE SIGNIFICANCE: The installation of a Type I or Type II exhaust hood is no longer required for a commercial cooking appliance that is provided with an integral downdraft exhaust system that is listed and labeled in accordance with NFPA 96, *Standard for Ventilation Control, and Fire Protection Cooking Operations.* Because these appliances have a built-in exhaust system, requiring a hood over the appliance serves no purpose. This application is common in food establishments where the cooking is done in front of the customer directly at the table.

Hibachi table with integral downdraft exhaust system *(Courtesy of Roaster Tech)*

507.2.1

Type I Hoods

CHANGE TYPE: Modification

CHANGE SUMMARY: Type I hoods no longer are required to be installed where complying electric cooking appliances are being used.

2012 CODE: 507.2.1 Type I Hoods. Type I hoods shall be installed where cooking appliances produce grease or smoke as a result of the cooking process. Type I hoods shall be installed over medium-duty, heavy-duty and extra-heavy-duty cooking appliances. Type I hoods shall be installed over light-duty cooking appliances that produce grease or smoke.

> **Exception:** A Type I hood shall not be required for an electric cooking appliance where an approved testing agency provides documentation that the appliance effluent contains 5 mg/m^3 or less of grease when tested at an exhaust flow rate of 500 cfm (0.236 m^3/s) in accordance with Section 17 of UL 710B.

CHANGE SIGNIFICANCE: Where the cooking process does not produce quantities of grease exceeding the prescribed threshold, a Type I hood is no longer required for electric cooking appliances. The IMC provisions are now current with those of NFPA 96, *Standard for Ventilation Control, and Fire Protection Cooking Operations,* and UL 710 B, *Recirculating Systems*, that allow for the elimination of a Type I hood where grease emissions are minimal or nonexistent.

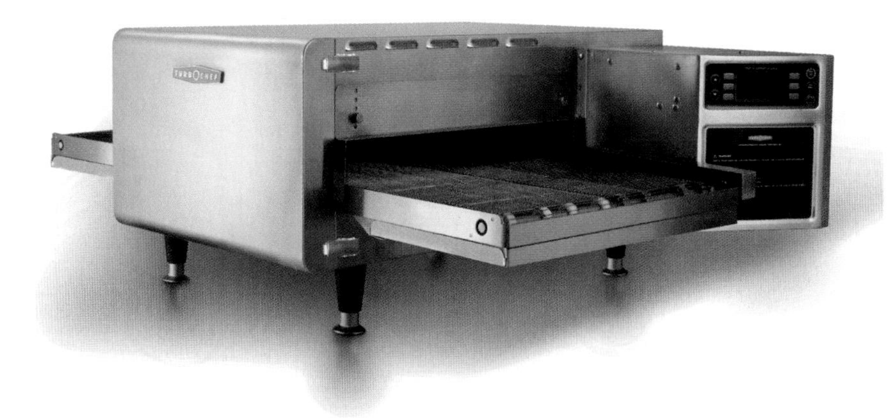

Electric oven where a hood is not required *(Courtesy of TurboChef Global)*

507.2.1.1

Operation of Type I Hoods

CHANGE TYPE: Modification

CHANGE SUMMARY: A method is now required to keep the pilot burner on a gas cooking appliance from being extinguished when the kitchen exhaust fan interlock shuts off appliances.

2012 CODE: **507.2.1.1 Operation.** Type I hood systems shall be designed and installed to automatically activate the exhaust fan whenever cooking operations occur. The activation of the exhaust fan shall occur through an interlock with the cooking appliances, by means of heat sensors or by means of other approved methods. A method of interlock between an exhaust hood system and appliances equipped with standing pilot burners shall not cause the pilot burners to be extinguished. A method of interlock between an exhaust hood system and cooking appliances shall not involve or depend upon any component of a fire extinguishing system.

CHANGE SIGNIFICANCE: The addition to this section in the IMC that deals with standing pilot burners was taken from Section 505.1.1 in the IFGC in order to make it easier for the user to find the necessary information and avoid a possible oversight. If a device is used in the interlock system between the exhaust fan on a Type I hood and a gas cooking appliance that will shut off the gas supply to the cooking appliance, the shut-off device shall not cause any pilot burner to be extinguished. This requirement will

507.2.1.1 continues

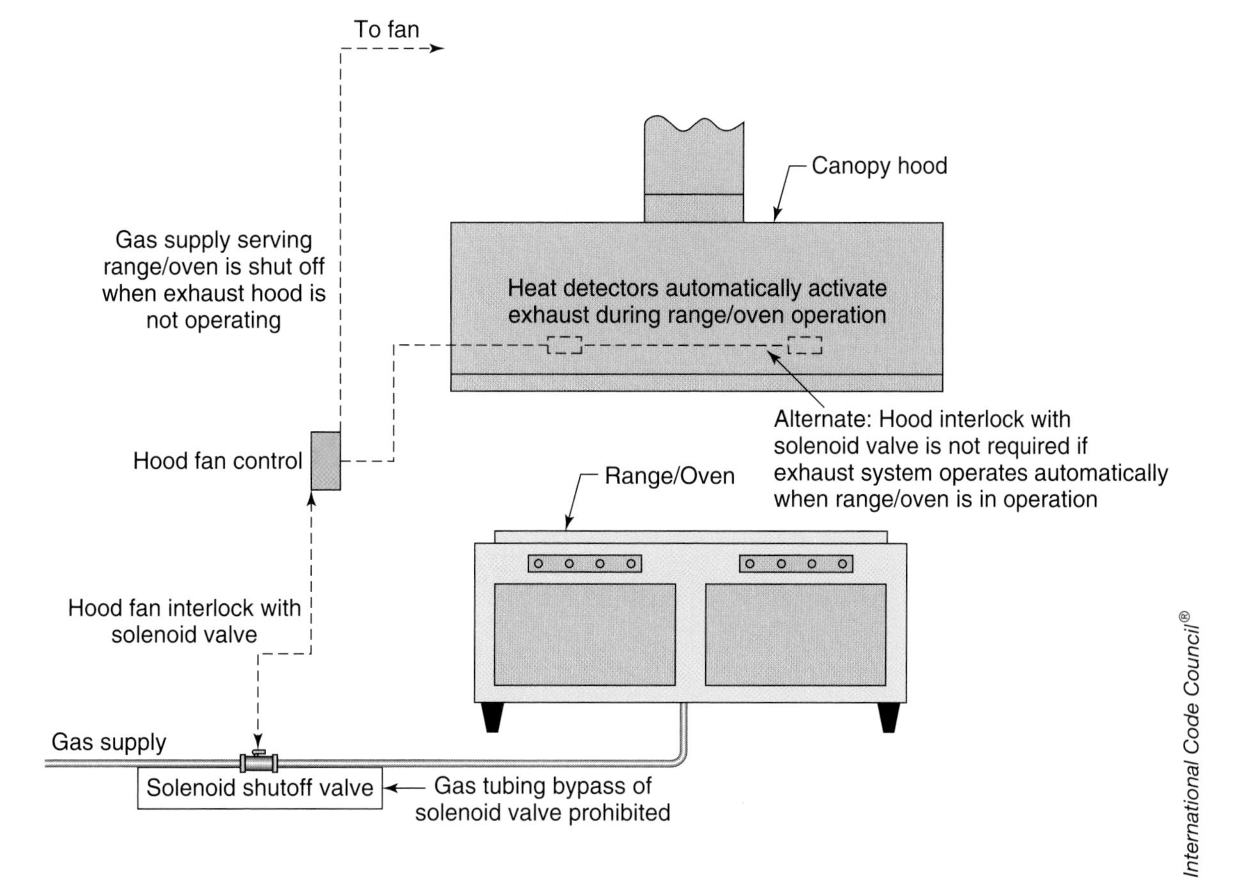

Commercial cooking appliance vented by exhaust hoods

507.2.1.1 continued

prevent the inconvenience of having to relight the pilot burner every time the Type I exhaust system is turned off.

An additional change addresses the practice of interlocking and connecting to a shut-off device that is a component of a fire extinguishing system in order to comply with general provisions of Section 507.2.1.1. This could pose a potential problem for the shut-off device because it is not listed to provide this function, which might also compromise the effectiveness of the fire extinguishing system in an emergency situation.

507.2.1.2

Exhaust Flow Rate Label for Type I Hoods

CHANGE TYPE: Addition

CHANGE SUMMARY: Manufacturers of listed Type I commercial cooking hoods are now required to provide information on a label attached to the hood specifying the listed minimum exhaust air flow for the hood based upon the cooking appliance duty classification.

2012 CODE: **507.2.1.2 Exhaust Flow Rate Label.** Type I hoods shall bear a label indicating the minimum exhaust flow rate in CFM per linear foot (1.55 L/s per linear meter) of hood that provides for capture and containment of the exhaust effluent for the cooking appliances served by the hood, based on the cooking appliance duty classifications defined in this code.

CHANGE SIGNIFICANCE: The manufacturers of listed Type I commercial cooking hoods must now provide a label on the hood specifying what exhaust flow rate the hood was tested to and listed for based on the cooking appliance duty classifications. Historically, hoods that have been listed to UL Standard 710 are tested and listed for a certain amount of exhaust air flow per linear foot of hood based on the amount of heat the cooking appliance the hood is serving will generate. The temperature rating of each cooking appliance is not always available to an inspector, and compliance with the code is often difficult to determine. Cooking appliance duties are defined as light, medium, heavy, and extra-heavy duty. The label is now required to base the exhaust flow rate on the cooking appliance duty classification that is defined in the code.

LISTING DESCRIPTION
TESTED, LISTED, AND APPROVED TO EXHAUST A MINIMUM OF 200 CFM PER LINEAR FOOT OVER 600-DEGREE COOKING EQUIPMENT

International Code Council®

Type I hood label

507.2.2

Type II Hoods

CHANGE TYPE: Modification

CHANGE SUMMARY: A Type II hood is now required to be installed above all appliances that produce products of combustion but do not produce grease or smoke. An exact exhaust rate is specified for areas where a cooking appliance is being used but a Type II hood is not required.

2012 CODE: **507.2.2 Type II Hoods.** Type II hoods shall be installed above dishwashers and appliances that produce heat or moisture and do not produce grease or smoke as a result of the cooking process, except where the heat or moisture loads from such appliances are incorporated into the HVAC system design or into the design of a separate removal system. Type II hoods shall be installed above all light duty appliances that produce products of combustion and do not produce grease or smoke as a result of the cooking process. Spaces containing cooking appliances that do not require Type II hoods shall be ~~ventilated~~ provided with exhaust at a rate of 0.70 cfm per square foot (0.00033 m^3/s). ~~in accordance with Section 403.3.~~ For the purpose of determining the floor area required to be ~~ventilated~~ exhausted, each individual appliance that is not required to be installed under a Type II hood shall be considered as occupying not less than 100 square feet. Such additional square footage shall be provided with exhaust at a rate of 0.70 cfm per square foot.

CHANGE SIGNIFICANCE: Previously, a space or area where a cooking appliance was allowed to operate without a Type II hood was required to be ventilated in accordance with Section 403.3 in the IMC. Table 403.3 in the IMC does not establish any values for outside air in a kitchen, as it only specifies an exhaust rate. For clarity purposes, the exhaust rate of 0.70 cfm per square foot taken from Table 403.3 refers to a space where a cooking appliance is being used without a Type II hood. A Type II hood is now permitted to be used with appliances that are rated for other than light duty and do not produce grease, smoke, or combustion products.

The addition of the text "as a result of the cooking process" is intended to clarify that the smoke being referenced is that smoke produced as part of the normal cooking process and not a result of the food being burned. As an example, toast that is burned in a toaster and produces smoke would not establish the need for a Type II hood.

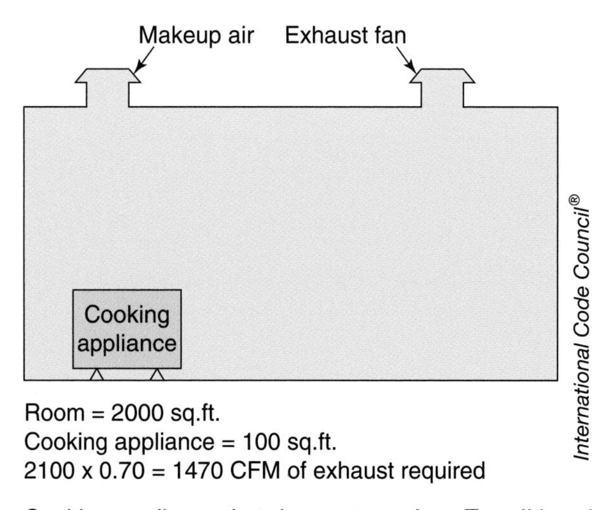

Cooking appliance that does not require a Type II hood

507.10

Hoods Penetrating a Ceiling

CHANGE TYPE: Addition

CHANGE SUMMARY: Field-applied grease duct enclosure systems are now specifically prohibited from being used as enclosures over the top of Type I hoods.

2012 CODE: **507.10 Hoods Penetrating a Ceiling.** Type I hoods or portions thereof penetrating a ceiling, wall or furred space shall comply with Section 506.3.11. Field-applied grease duct enclosure systems, as addressed in Section 506.3.11.2, shall not be utilized to satisfy the requirements of this section.

CHANGE SIGNIFICANCE: Field-applied grease duct enclosure systems are allowed to be used to enclose commercial kitchen grease ducts based on the provisions in Section 506.3.11.2. The listing for such duct enclosure systems specifies that they have been tested for enclosing a grease duct only. Code enforcement personnel have encountered numerous occasions where the enclosure system material has been used as an enclosure around a Type I hood that penetrates a ceiling and is required to be enclosed as required for a grease duct in Section 506.3.11 in the IMC. Section 507.10 now clearly states that field-applied grease duct enclosure systems cannot be used to satisfy the requirements for the protection of Type I hoods.

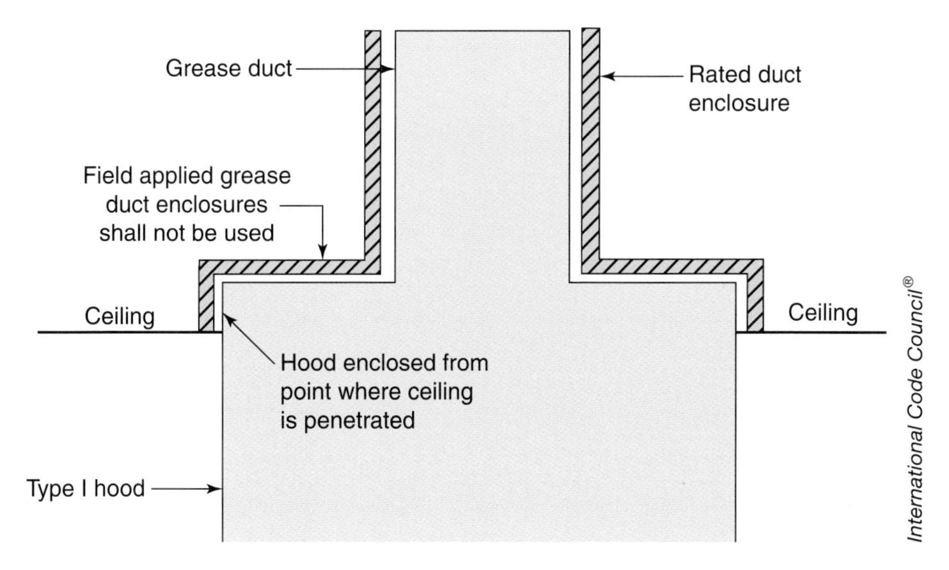

Hood enclosure

510.7

Fire Suppression Required for Hazardous Exhaust Ducts

CHANGE TYPE: Modification

CHANGE SUMMARY: Automatic fire suppression systems are no longer required in the exhaust ducts in semiconductor fabrication facilities.

2012 CODE: **510.7 Suppression Required.** Ducts shall be protected with an approved automatic fire suppression system installed in accordance with the *International Building Code.*

Exceptions:

1. An approved automatic fire suppression system shall not be required in ducts conveying materials, fumes, mists and vapors that are nonflammable and noncombustible under all conditions and at any concentrations.
2. Automatic fire suppression systems shall not be required in metallic and noncombustible nonmetallic exhaust ducts in semiconductor fabrication facilities.

~~2.~~ 3. An approved automatic fire suppression system shall not be required in ducts where the largest cross-sectional diameter of the duct is less than 10 inches (254 mm).

~~3.~~ 4. For laboratories, as defined in Section 510.1, automatic fire protection systems shall not be required in laboratory hoods or exhaust systems.

CHANGE SIGNIFICANCE: A conflict between the IFC and the IMC has been eliminated regarding the exemption of specific types of ducts in Group H-5 occupancies from the requirement for fire suppression. A new exception has been added to IMC Section 510.7 to specifically address semiconductor fabrication facilities so that other occupancies are not affected.

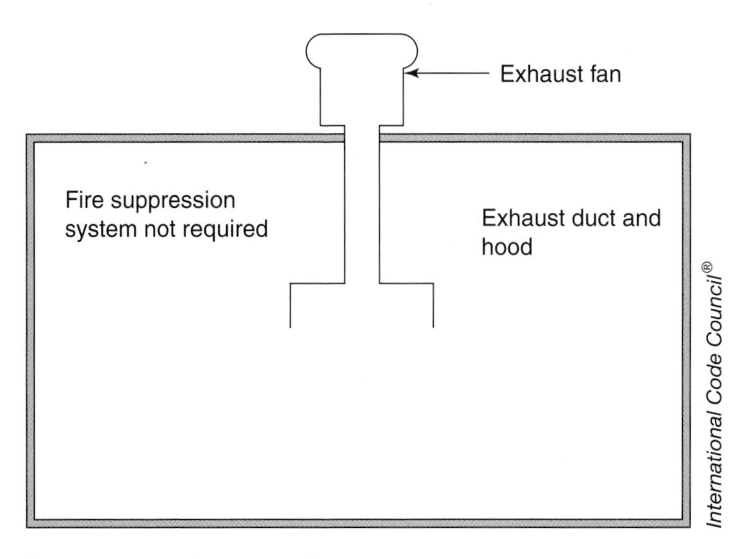

Semiconductor fabrication facility

601.4

Contamination Prevention in Plenums

CHANGE TYPE: Modification

CHANGE SUMMARY: Chimneys and vents are now permitted to pass through a plenum where in compliance with one of three new allowances.

2012 CODE: 601.4 Contamination Prevention. Exhaust ducts under positive pressure, chimneys and vents shall not extend into or pass through ducts or plenums.

Exceptions:

1. Exhaust systems located in ceiling return air plenums over spaces that are permitted to have 10 percent recirculation in accordance with Section 403.2.1, Item 4. The exhaust duct joints, seams, and connections shall comply with Section 603.9.
2. This section shall not apply to chimneys and vents that pass through plenums where such venting systems comply with one of the following requirements:
   - 2.1. The venting system shall be listed for positive pressure applications and shall be sealed in accordance with the vent manufacturer's instructions.
   - 2.2. The venting system shall be installed such that fittings and joints between sections are not installed in the above ceiling space.

601.4 continues

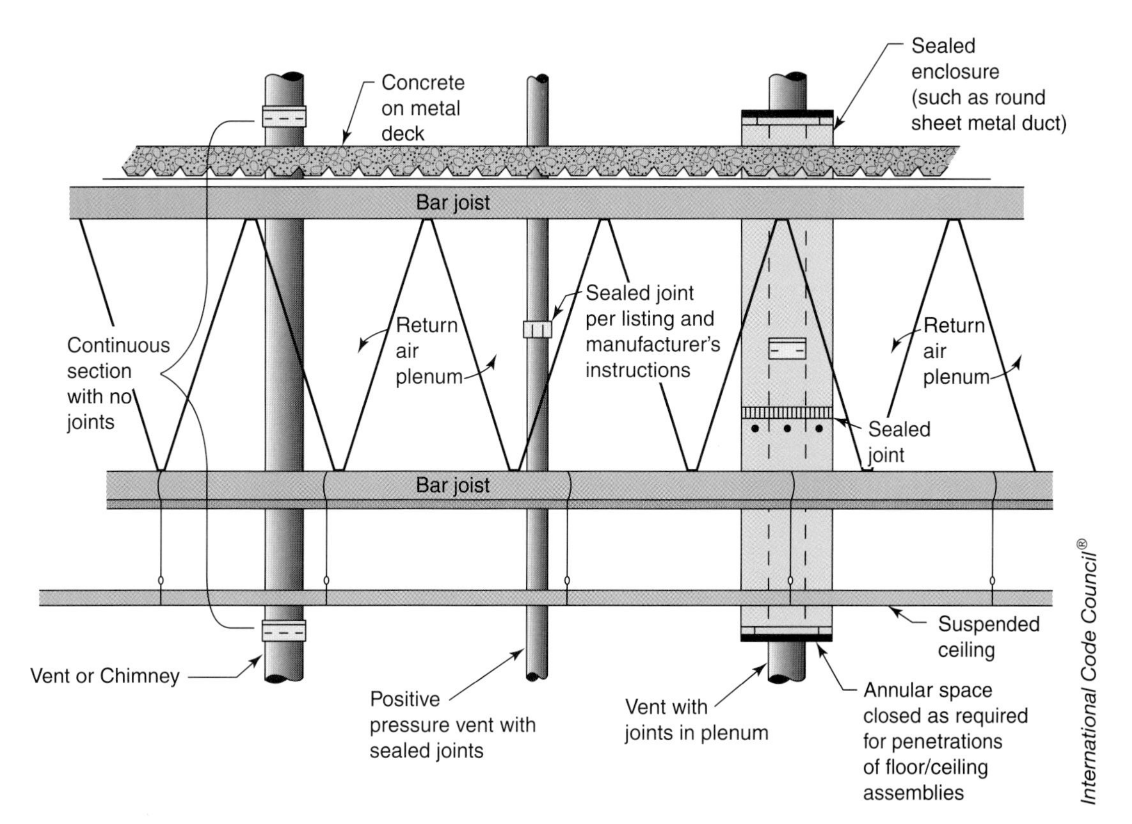

Chimney or vent in a plenum

601.4 continued

2.3. The venting system shall be installed in a conduit or enclosure with sealed joints separating the interior of the conduit or enclosure from the ceiling space.

CHANGE SIGNIFICANCE: Chimneys and vents are now permitted to pass through a plenum under certain conditions. The need for such an arrangement usually occurs where there is an open return air plenum above a ceiling in a commercial building and a vent or chimney from a fuel-burning appliance must pass through the plenum in order to terminate to the exterior of the building. The conditions of the new exception provide for a means to prevent flue gases from a chimney or vent from escaping and entering into the return air plenum and circulating throughout the building.

602.2.1
Materials within Plenums

CHANGE TYPE: Clarification

CHANGE SUMMARY: It has been clarified that any material or assembly that encloses a combustible material in a plenum must be noncombustible, gypsum board, or listed and labeled as part of a tested assembly or system.

2012 CODE: 602.2.1 Materials within Plenums. Except as required by Sections 602.2.1.1 through ~~602.2.1.6~~ 602.2.1.5 materials within plenums shall be noncombustible or shall be listed and labeled as having a flame spread index of not more than 25 and a smoke-developed index of not more than 50 when tested in accordance with ASTM E 84 or UL 723.

Exceptions:

1. Rigid and flexible ducts and connectors shall conform to Section 603.
2. Duct coverings, linings, tape and connectors shall conform to Sections 603 and 604.
3. This section shall not apply to materials exposed within plenums in one- and two-family dwellings.
4. This section shall not apply to smoke detectors.
5. Combustible materials fully enclosed within one of the following:
   - 5.1. Continuous noncombustible raceways or enclosures
   - 5.2. Approved gypsum board assemblies
   - 5.3. ~~Within~~ Materials listed and labeled for ~~such application~~ installation within a plenum.

602.2.1 continues

Enclosure shall be noncombustible, approved gypsum board assembly, or shall be listed and labeled for installation in a plenum

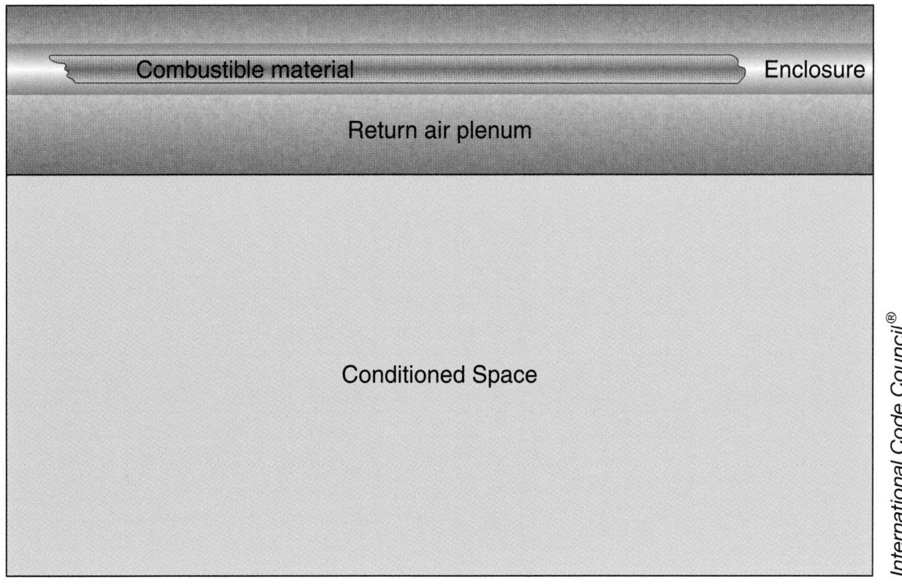

International Code Council®

Material within a plenum

602.2.1 continued

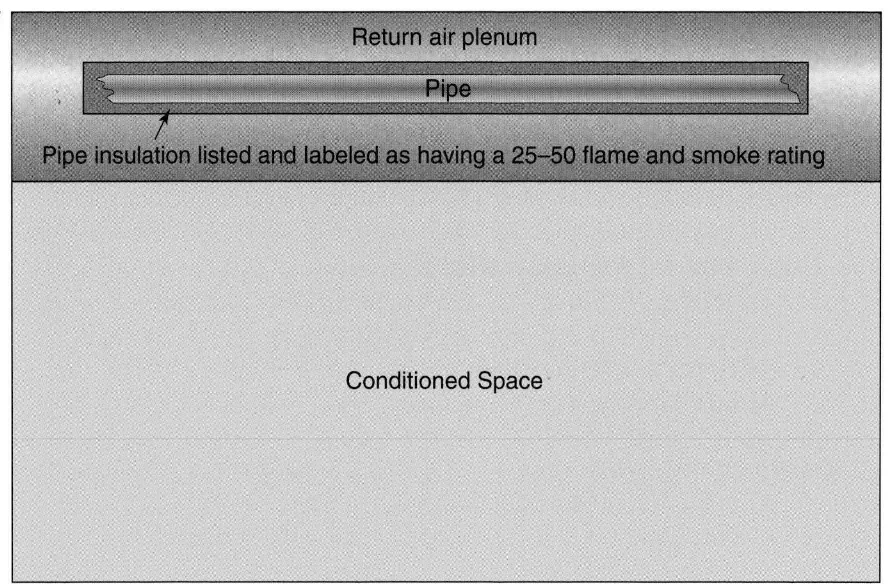

All material within a plenum shall be noncombustible or shall be listed and labeled as having a flame spread rating of not more than 25 and a smoke-developed index of not more than 50 when tested in accordance with ASTM E-84 or UL 728

6. Materials in Group H, Division 5 fabrication areas and the areas above and below the fabrication area that share a common air recirculation path with the fabrication area.

CHANGE SIGNIFICANCE: The enclosure methods previously established in Exception 5 have been reformatted so they are easier for the user of the code to identify. Item 5.3 clarifies that if a combustible material located in a plenum is fully enclosed, the system or enclosure material must be noncombustible, be an approved gypsum board assembly, or be listed and labeled for installation within a plenum. In addition, the provisions addressing semiconductor fabrication areas previously located in Section 602.2.1.6 have been relocated to Section 602.2.1 as Exception 6.

603.7

Rigid Duct Penetrations

CHANGE TYPE: Modification

CHANGE SUMMARY: In relationship to the required garage/dwelling separation, only those ducts that penetrate a wall or ceiling between the dwelling and the adjacent private garage need comply with Section 603.7.

2012 CODE: 603.7 Rigid Duct Penetrations. Duct system penetrations of walls, floors, ceilings and roofs and air transfer openings in such building components shall be protected as required by Section 607. Ducts in a private garage ~~and ducts~~ that penetrate~~ing~~ ~~the~~ a wall~~s~~ or ceiling~~s~~ that ~~separating~~ separates a dwelling from a private garage shall be continuous, ~~and~~ shall be constructed of sheet steel having a minimum thickness of ~~26 gage~~ 0.0187 inch (0.4712 mm) (No. 26 gage) ~~galvanized sheet metal~~ and shall not have openings into the garage. Fire and smoke dampers are not required in such ducts passing through the wall or ceiling separating a dwelling from a private garage except where required by Chapter 7 of the *International Building Code*.

CHANGE SIGNIFICANCE: In the 2009 IMC, Section 603.7 prohibited duct openings in a private garage in all installations, even where such ducts did not penetrate the garage/dwelling separation elements. The intention has always been not to allow openings in a duct system located in a private garage where the duct penetrates a wall or a ceiling between a private garage and a dwelling. Ducts that serve a private garage only and do not penetrate a wall or ceiling now are not required to be constructed of 26-gage metal and are permitted to have openings.

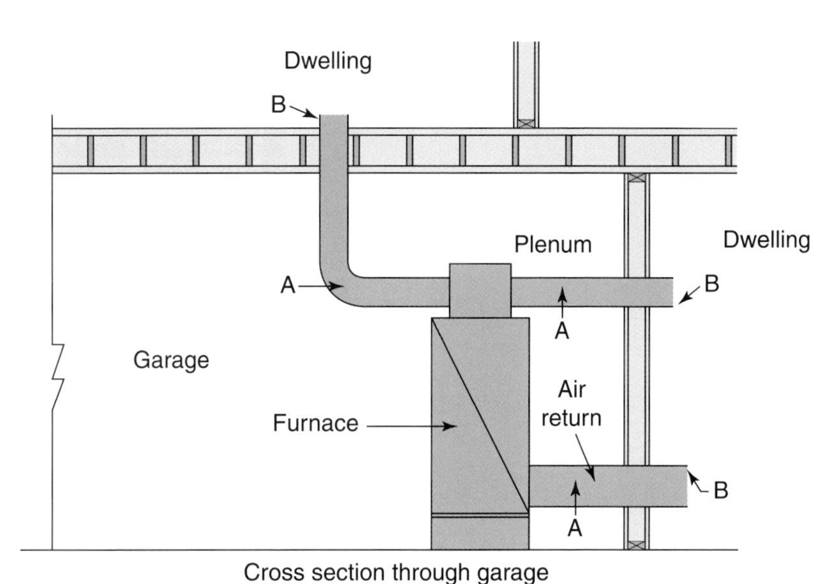

Ducts A - 0.019 inch (No. 26 gauge) galvanized steel with no openings into garage
Ducts B - Any duct approved by the Mechanical Code

Duct penetration of a garage/dwelling separation

603.9

Duct Joints, Seams, and Connections

CHANGE TYPE: Clarification

CHANGE SUMMARY: Unlisted duct tape is no longer permitted as a sealant on nonmetallic ducts.

2012 CODE: **603.9 Joints, Seams, and Connections.** All longitudal and transverse joints, seams and connections in metallic and nonmetallic ducts shall be constructed as specified in SMACNA HVAC Duct Construction Standards—Metal and Flexible and NAIMA Fibrous Glass Duct Construction Standards. All joints, longitudinal and transverse seams, and connections in ductwork shall be securely fastened and sealed with welds, gaskets, mastics (adhesives), mastic-plus-embedded- fabric systems, liquid sealants or tapes. Closure systems used to seal ductwork listed and labeled in accordance with UL 181A shall be marked "181A-P" for pressure-sensitive tape, "181 A-M" for mastic or "181 A-H" for heat-sensitive tape. Closure systems used to seal flexible air ducts and flexible air connectors shall comply with UL 181B and shall be marked "181B-FX" for pressure-sensitive tape or "181B-M" for mastic. Duct connections to flanges of air distribution system equipment shall be sealed and mechanically fastened. Mechanical fasteners for use with flexible nonmetallic air ducts shall comply with UL 181B and shall be marked "181B-C." Closure systems used to seal metal ductwork shall be installed in accordance with the manufacturer's installation instructions. Unlisted duct tape is not permitted as a sealant on any ~~metal~~ duct.

> **Exception:** Continuously welded and locking-type longitudinal joints and seams in ducts operating at static pressures less than 2 inches of water column (500 Pa) pressure classification shall not require additional closure systems.

CHANGE SIGNIFICANCE: It has been clarified that unlisted duct tape is not allowed as a sealant on any type of duct system. The previous limitation only applied to metal ducts.

Listed duct tape *(Courtesy of Venture Tape)*

603.17, 202

Air Dispersion Systems

CHANGE TYPE: Addition

CHANGE SUMMARY: Air dispersion systems as defined in Section 202 and recognized in UL 2518 are now permitted to be installed.

2012 CODE: **202 Air Dispersion System.** Any diffuser system designed to both convey air within a room, space or area and diffuse air into that space while operating under positive pressure. Systems are commonly constructed of, but not limited to, fabric or plastic film.

603.17. Air dispersion systems. Air dispersion systems shall:

1. Be installed entirely in exposed locations.
2. Be utilized in systems under positive pressure.
3. Not pass through or penetrate fire–resistant rated construction.
4. Be listed and labeled in compliance with UL 2518.

Chapter 15
UL

2518–02 Air Dispersion System Materials

CHANGE SIGNIFICANCE: A duct system that has not previously been recognized is now acceptable for installation. Air dispersion systems are listed to UL 2518, *Air Dispersion System Materials*, which has been added to Chapter 15.

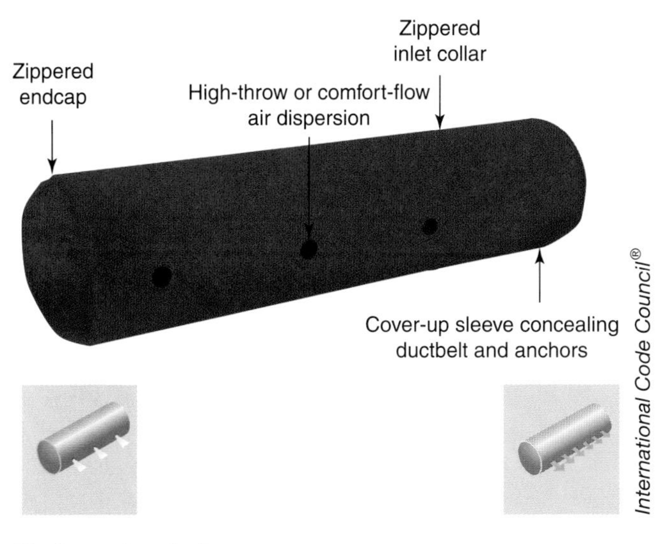

Air dispersion duct

805.3

Factory-Built Chimney Offsets

CHANGE TYPE: Addition

CHANGE SUMMARY: The maximum offset in a factory-built chimney is now specified and the number of offsets has been limited.

2012 CODE: **805.3 Factory Built Chimney Offsets.** Where a factory-built chimney assembly incorporates offsets, no part of the chimney shall be at an angle of more than 30 degrees from vertical at any point in the assembly and the chimney assembly shall not include more than 4 elbows.

CHANGE SIGNIFICANCE: UL 103 addressing factory-built chimneys is the basis for the limitations specified in the new Section 805.3. There has always been confusion about offsets in factory-built chimneys, and, because of the lack of specific code provisions, reference was commonly made to the offset requirements for Type B vents used with gas-fired appliances. The new requirements now specify the maximum permitted offset in a factory-built chimney, as well as the maximum number of offsets allowed.

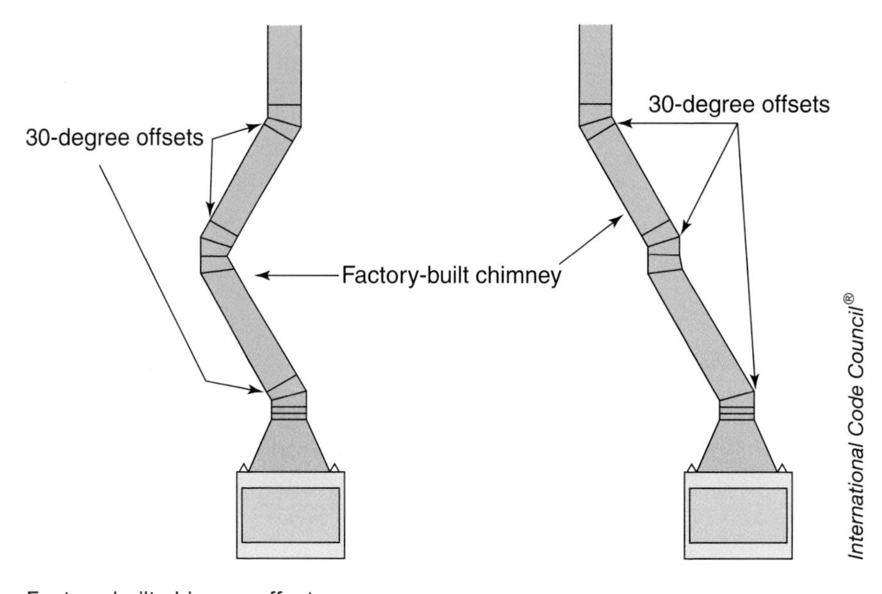

Factory-built chimney offset

901.4

Fireplace Accessories

CHANGE TYPE: Addition

CHANGE SUMMARY: Fireplace accessories must now comply with UL 907, which has been added to Chapter 15.

2012 CODE: 901.4 Fireplace Accessories. Listed and labeled fireplace accessories shall be installed in accordance with the conditions of the listing and the manufacturer's installation instructions. Fireplace accessories shall comply with UL 907.

Chapter 15
UL

907-94 Fireplace Accessories–with revisions through July 2006

CHANGE SIGNIFICANCE: Historically, listed fireplace accessories were required to be installed in accordance with the listing and the manufacturers' installation instructions. Now fireplace accessories must be listed and labeled to comply with UL 907, *Fireplace Accessories*. Among other items, UL 907 for fireplace accessories addresses glass door assemblies, combustion air vents, smoke chambers, surfacing materials, and termination caps.

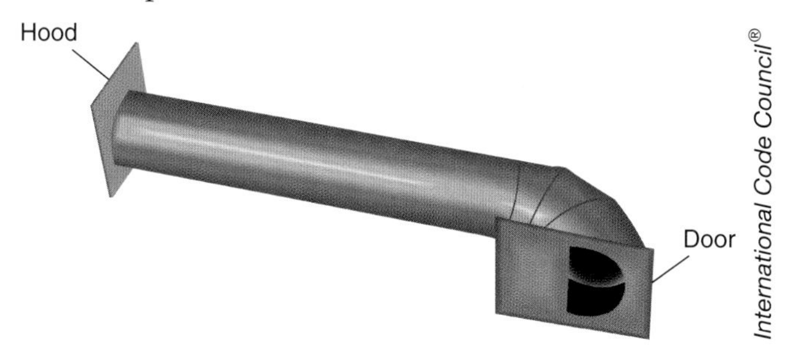

Listed outside air intake regulator for masonry fireplace

928

Evaporative Cooling Equipment

CHANGE TYPE: Addition

CHANGE SUMMARY: Requirements for the installation of evaporative coolers have been introduced into the IMC in the new Section 928.

2012 CODE:

SECTION 928
EVAPORATIVE COOLING EQUIPMENT

928.1 General. Evaporative cooling equipment shall:

1. Be installed in accordance with the manufacturer's installation instructions.
2. Be installed on level platforms in accordance with section 304.10.
3. Have openings in exterior walls or roofs flashed in accordance with the *International Building Code.*
4. Be provided with portable water backflow protection in accordance with Section 608 of the *International Plumbing Code.*
5. Have air intake opening locations in accordance with Section 401.4.

CHANGE SIGNIFICANCE: Evaporative coolers have not been addressed in previous editions of the IMC. Requirements now address such issues as backflow, air intake location openings in roofs and exterior walls, and manufacturers' installation instructions.

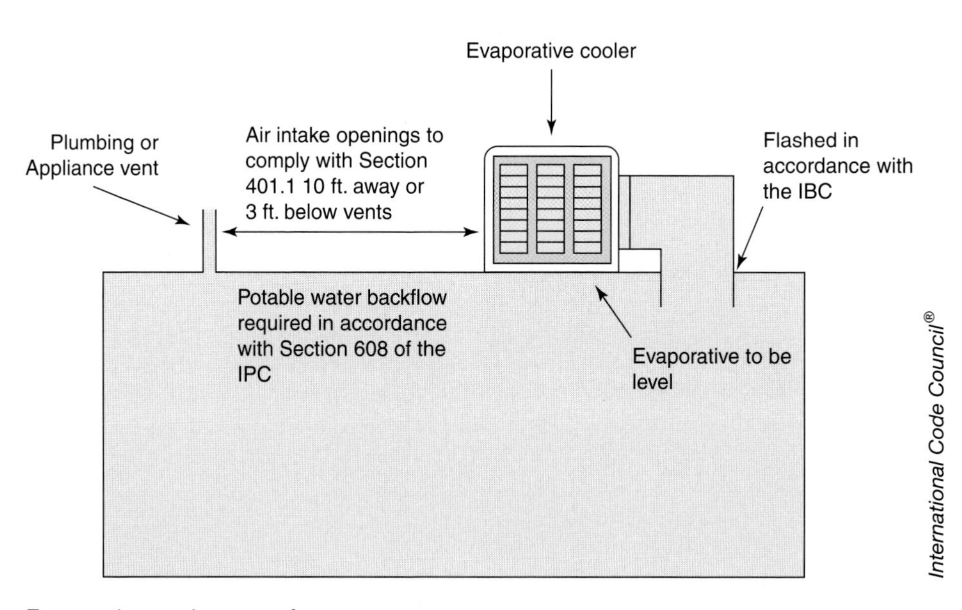

Evaporative cooler on roof

1101.10

Locking Access Port Caps

CHANGE TYPE: Modification

CHANGE SUMMARY: Locking caps are no longer required on refrigerant access ports if the refrigeration equipment is located in a secured location.

2012 CODE: 1101.10 Locking Access Port Caps. Refrigerant circuit access ports located outdoors shall be fitted with locking-type tamper-resistant caps or shall be otherwise secured to prevent unauthorized access.

CHANGE SIGNIFICANCE: Section 1101.10 requiring that locking caps be used to secure refrigerant access ports was added to the 2009 IMC to prevent huffing. There are other methods of securing these access ports other than locking caps. If the refrigeration equipment is located in an outdoor location that is not accessible to the public, the intention of the code has been met and thus the locking caps would be unnecessary.

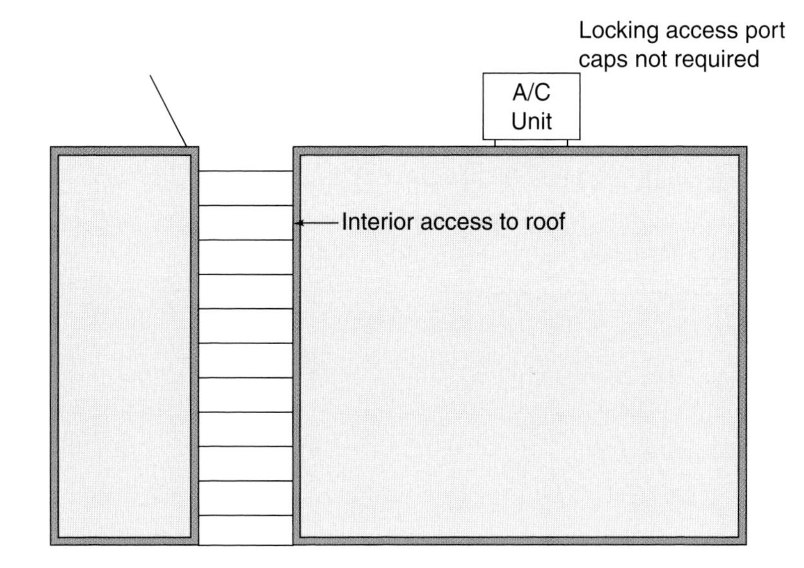

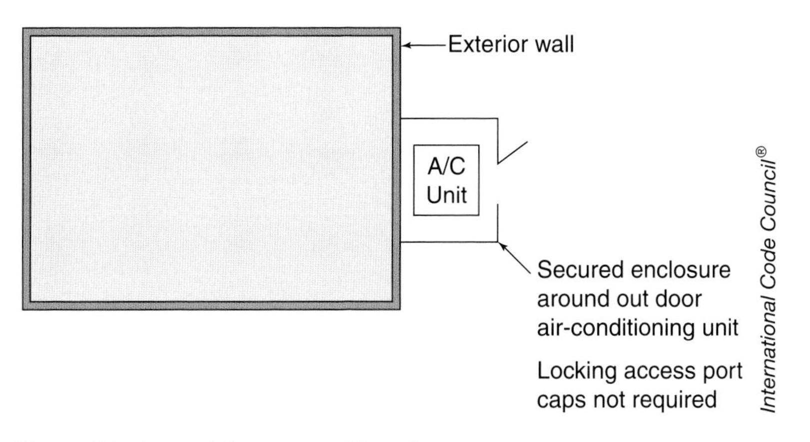

Air-conditioning unit in a secured location

1105.6, 1105.6.3

Machinery Room Ventilation

CHANGE TYPE: Modification

CHANGE SUMMARY: The minimum ventilation rates in an ammonia machinery room must now be in accordance with IIAR2.

2012 CODE: 1105.6 Ventilation. Machinery rooms shall be mechanically ventilated to the outdoors. ~~Mechanical ventilation shall be capable of exhausting the minimum quality of air both at the normal operating and emergency conditions. Multiple fans or multispeed fans shall be allowed in order to produce the emergency ventilation rate and to obtain a reduced airflow for normal ventilation.~~

Exception: (no changes to text)

1105.6.3 Ventilation Rate. For the other than ammonia systems, the mechanical ventilation systems shall be capable of exhausting the minimum quantity of air both at normal operating and emergency conditions, as required by Sections 1105.6.3.1 and 1105.6.3.2. The minimum required ventilation rate for ammonia shall be 30 air changes per hour, in accordance with IIAR2. Multiple fans or multispeed fans shall be allowed to produce the emergency ventilation rate and to obtain a reduced airflow for normal ventilation.

CHANGE SIGNIFICANCE: Ventilation rates in ammonia machinery rooms will now be based on IIAR2 requirements, making the IMC current with industry standards.

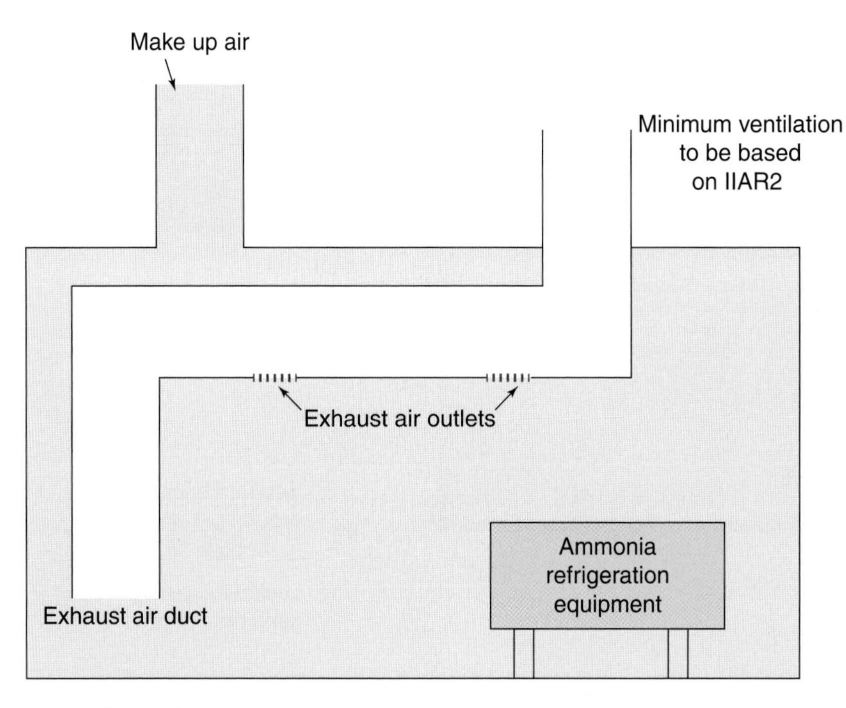

Ammonia machinery room

International Code Council®

1106.4

Flammable Refrigerants

CHANGE TYPE: Modification

CHANGE SUMMARY: The ventilation requirements of Section 1106.3 for ammonia machinery rooms are now mandatory in order to be exempted from the Class 1, Division 2 hazardous location requirements of NFPA 70.

2012 CODE: 1106.4 Flammable Refrigerants. Where refrigerants of Groups A2, A3, B2 and B3 are used, the machinery room shall conform to the Class 1, Division 2 hazardous location classification requirements of NFPA 70.

> **Exception:** Ammonia machinery rooms that are provided with ventilation in accordance with Section 1106.3.

CHANGE SIGNIFICANCE: In order to create consistency with the *International Fire Code*, the exception to Section 1106.4 now specifies that an ammonia machinery room must be ventilated to qualify for the exception that would exempt the Class 1, Division 2 hazardous location classification requirements of NFPA 70, *National Electrical Code*.

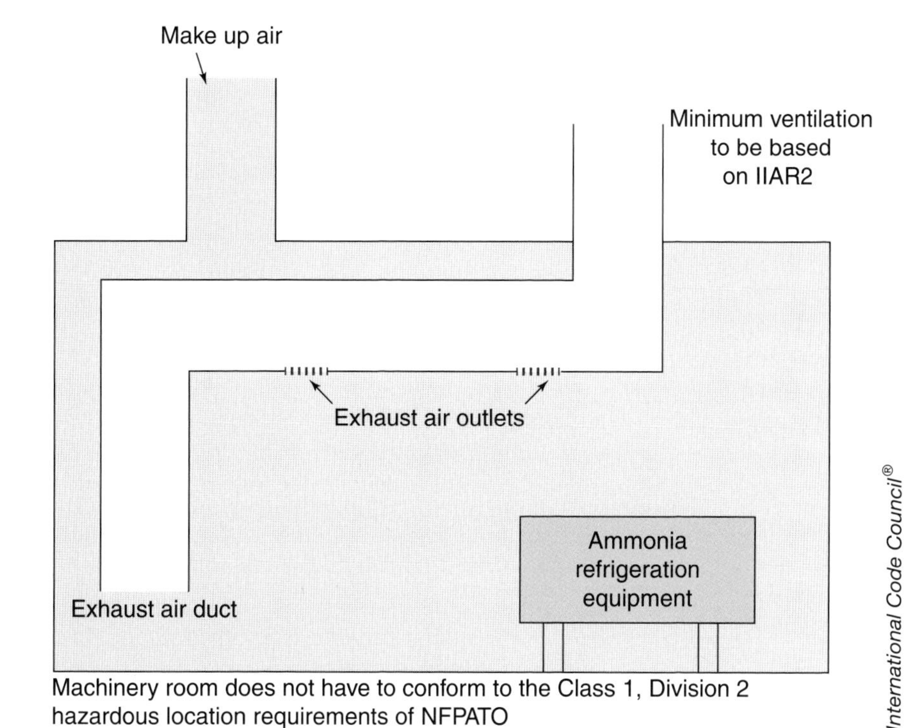

Ammonia machinery room

PART 3

International Fuel Gas Code

Chapters 1 through 8

- **Chapter 1** Administration No Changes Addressed
- **Chapter 2** Definitions
- **Chapter 3** General Regulations
- **Chapter 4** Gas Piping Installations
- **Chapter 5** Chimneys and Vents No Changes Addressed
- **Chapter 6** Specific Appliances
- **Chapter 7** Gaseous Hydrogen Systems No Changes Addressed
- **Chapter 8** Referenced Standards No Changes Addressed

The *International Fuel Gas Code* (IFGC) applies to the installation of fuel gas piping systems, fuel gas utilization equipment, gaseous hydrogen systems, and related accessories. Chapter 1 provides for the administration and enforcement of the code, assigning responsibility and authority to the code official. Chapter 2 contains definitions of terms specific to their use throughout the code. The general requirement provisions of Chapter 3 govern the approval and installation of all equipment and appliances regulated by the code. Requirements for the design and installation of gas piping systems are set out in Chapter 4 and include provisions for materials, components, fabrication, testing, inspection, operation, and maintenance of such systems. The scope of Chapter 5 includes factory-built chimneys, liners, vents, connectors, and masonry chimneys serving gas-fired appliances. Reference is made to the *International Mechanical Code* for chimneys serving appliances using other fuels and to the *International Building Code* for the construction requirements of masonry chimneys. Approval, design, and installation of specific appliances such as furnaces, boilers, water heaters, fireplaces, decorative appliances, room heaters, and clothes dryers are covered in Chapter 6. Chapter 7 covers the developing technology of gaseous hydrogen systems, including hydrogen generation and refueling operations, and provides reference to the applicable provisions of the *International Fire Code*. Chapter 8 provides a complete list of standards referenced in various sections of the code. ■

202, 401, 401.9, 404.10, 404.1

Identification, Testing and Certification

308.1

Clearance to Combustible Materials

404.2

CSST Piping Systems

404.18

Prohibited Devices

408.4

Sediment Trap

410.4

Excess Flow Valves

410.5, 202

Flashback Arrestor Check Valve

618.4

Prohibited Sources

202, 401.9, 401.10, 404.1

Identification, Testing and Certification

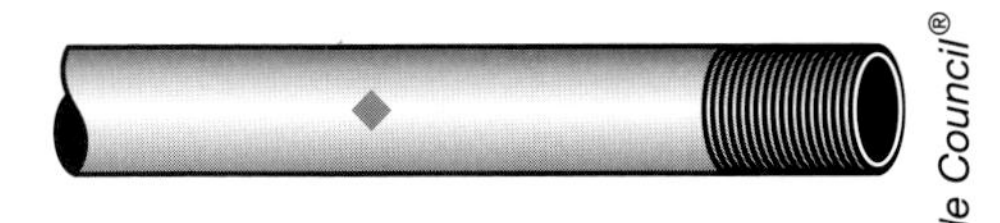

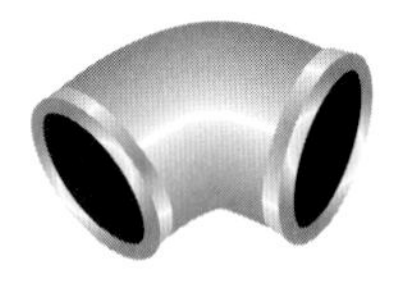

International Code Council®

Each length of pipe and fittings shall bear the identification of the manufacturer

CHANGE TYPE: Addition

CHANGE SUMMARY: Each section of pipe and each fitting utilized in a gas system requires the identification of the manufacturer.

2012 CODE:

SECTION 202
DEFINITIONS

Third-Party Certification Agency. An approved agency operating a product or material certification system that incorporates initial testing, assessment and surveillance of a manufacturer's quality control system.

Third-Party Certified. Certification obtained by the manufacturer indicating that the function and performance characteristics of a product or material have been determined by testing and ongoing surveillance by an approved third-party certification agency. Assertion of certification is in the form of identification in accordance with the requirements of the third-party certification agency.

Third-Party Tested. Procedure by which an approved testing laboratory provides documentation that a product, material or system conforms to specified requirements.

401.9 Identification. Each length of pipe and tubing and each pipe fitting, utilized in a fuel gas system shall bear the identification of the manufacturer.

401.10 Third-Party Testing and Certification. All piping, tubing and fittings shall comply with the applicable referenced standards, specifications and performance criteria of this code and shall be identified in accordance with Section 401.9. Piping, tubing and fittings shall either be tested by an approved third-party testing agency or certified by an approved third-party certification agency.

404.1 Installation of Materials. All materials used shall be installed in strict accordance with the standards under which the materials are accepted and approved. In the absence of such installation instructions shall be followed. Where the requirements of referenced standards or manufacturer's installation instructions do not conform to minimum provisions of this code, the provisions of this code shall apply.

CHANGE SIGNIFICANCE: Each section of pipe, tubing, and each fitting in a gas piping system must bear the identification of the manufacturer. This is the industry method of verifying that the pipe and fittings comply with the standards recognized in the code. Provisions have also been added to provide guidance on third-party certification.

308.1

Clearance to Combustible Materials

CHANGE TYPE: Clarification

CHANGE SUMMARY: It has been clarified that gypsum board is to be considered a combustible material for the purpose of required clearances, including those provisions of Section 308 addressing reductions in required clearances.

2012 CODE: **308.1 Scope.** This section shall govern the reduction in required clearances to combustible materials, including gypsum board, and combustible assemblies for chimneys, vents, appliances, devices and equipment. Clearance requirements for air-conditioning equipment and central heating boilers and furnaces shall comply with Sections 308.3 and 308.4.

CHANGE SIGNIFICANCE: Gypsum board, also known as drywall, is often incorrectly thought to be a noncombustible material. Because of this misunderstanding, appliances that require a specified clearance to combustible material have not been required to maintain the required clearance to gypsum board. It has now clarified that gypsum board is to be addressed in a manner consistent with any other combustible material when reducing clearances to combustibles.

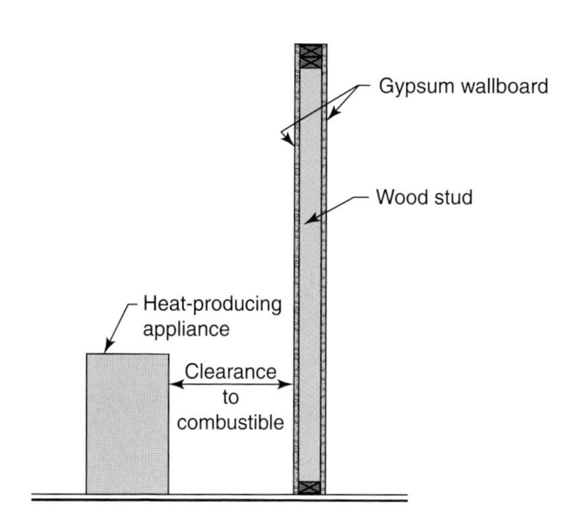

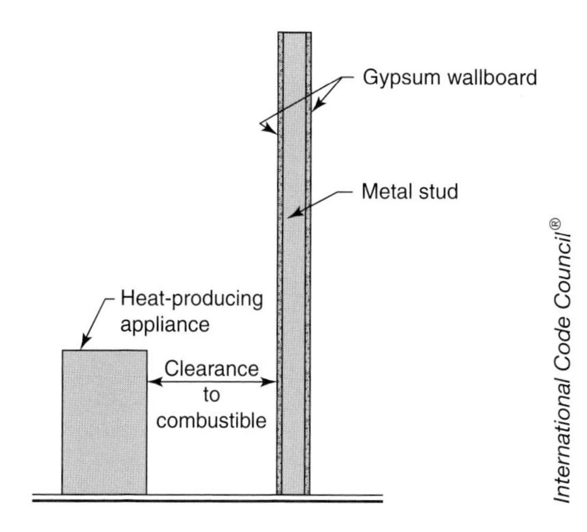

International Code Council®

Clearance to gypsum wallboard

404.2

CSST Piping Systems

CHANGE TYPE: Addition

CHANGE SUMMARY: CSST piping systems shall be installed in accordance with their listing and the manufacturer's installation instructions.

2012 CODE: **404.2 CSST.** CSST piping systems shall be installed in accordance with the terms of their approval, the conditions of listing, the manufacturer's installation instructions and this code.

CHANGE SIGNIFICANCE: CSST gas piping systems are listed to an ANSI standard and, like other equipment and appliances, are now required to be installed in accordance with the terms of their approval, the conditions of listing, and the manufacturer's installation instructions.

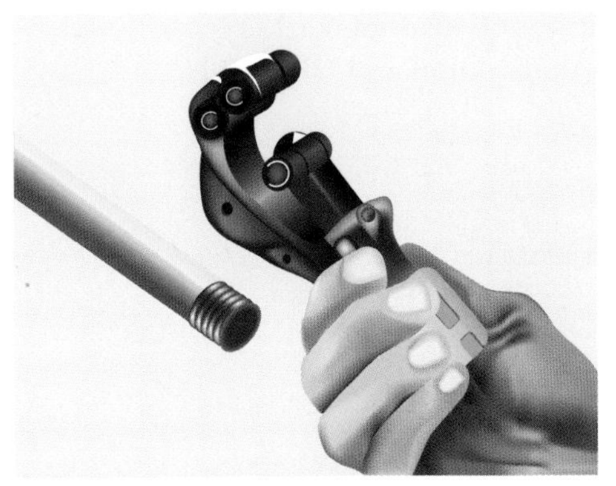

Step 1 – Cut tubing and remove PE coating to expose a minimum of four corrugations.

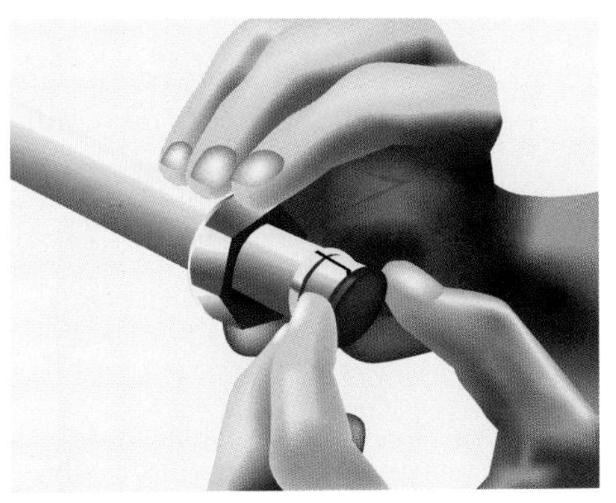

Step 2 – Slide nut over tubing and place retainer ring. Leave one corrugation exposed on the end of tubing.

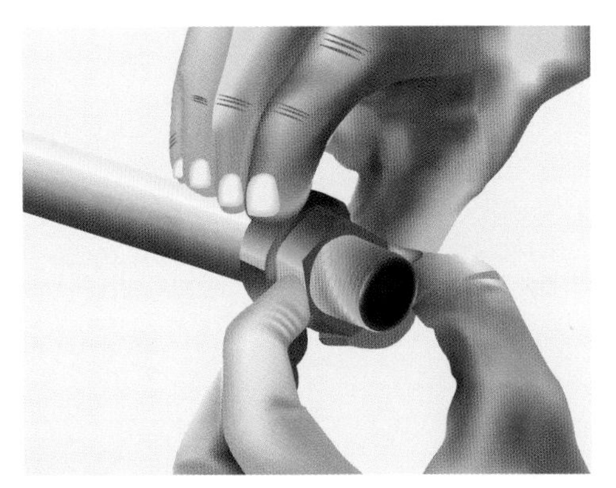

Step 3 – Slide nut over retainer and hand-tighten nut to body.

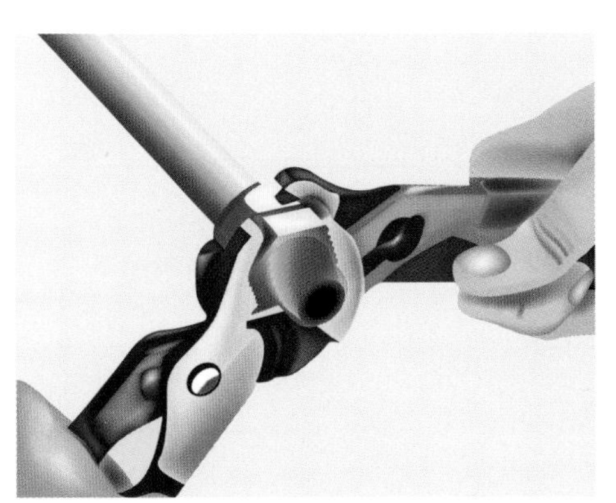

Step 4 – Tighten with wrenches until nut contacts body.

International Code Council®

Sample installation instructions for CSST gas piping

404.18

Prohibited Devices

CHANGE TYPE: Clarification

CHANGE SUMMARY: Excess flow valves and similar devices are now permitted to be placed in gas piping systems that have been sized to accommodate the pressure drop.

2012 CODE: **~~404.16~~ 404.18 Prohibited Devices.** A device shall not be placed inside the piping or fittings that will reduce the cross-sectional area or otherwise obstruct the free flow of gas.

Exceptions:

1. Approved gas filters.
2. An approved fitting or device where the gas piping system has been sized to accommodate the pressure drop of the fitting or device.

Listed excess flow valve

International Code Council®

CHANGE SIGNIFICANCE: Devices such as excess flow valves and earthquake valves are considered restrictions in gas piping systems and generally are prohibited from being installed in gas piping and fittings. It has now been clarified that if the gas piping system has been sized to accommodate a fitting or device placed in the gas piping system, the fitting or device is allowed to be installed.

408.4

Sediment Traps

CHANGE TYPE: Modification

CHANGE SUMMARY: An illustration of a sediment trap is now included within the IFGC in order to clarify the intent of the provisions.

2012 CODE: **408.4 Sediment Trap.** Where a sediment trap is not incorporated as part of the appliance, a sediment trap shall be installed downstream of the appliance shutoff valve as close to the inlet of the appliance as practical. The sediment trap shall be either a tee fitting having a capped nipple of any length installed vertically in the bottommost opening of the tee as illustrated in Figure 408.4 or other device approved as an effective sediment trap. Illuminating appliances, ranges, clothes dryers, decorative vented appliances for installation in vented fireplaces, gas fireplaces and outdoor grills need not be so equipped.

CHANGE SIGNIFICANCE: Because of misunderstandings as to how a sediment trap is to be constructed, an illustration has been added to the code text to clarify the intended installation method. In addition, decorative vented gas appliances and gas fireplaces are no longer required to be installed with a sediment trap. Although these appliances are susceptible to harm from debris, they are also manually operated; therefore, the user would be in attendance and aware of any problem.

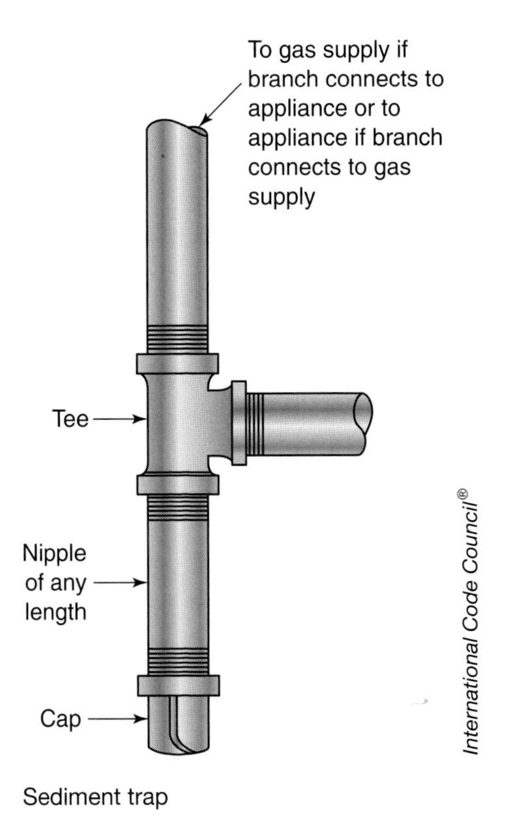

Sediment trap

410.4

Excess Flow Valves

CHANGE TYPE: Addition

CHANGE SUMMARY: An excess flow valve must now be listed, sized, and installed in accordance with the manufacturer's instructions.

2012 CODE: **410.4 Excess Flow Valves.** Where automatic excess flow valves are installed, they shall be listed for the application and shall be sized and installed in accordance with the manufacturers' instructions.

CHANGE SIGNIFICANCE: Excess flow valves are now required to be listed. In order to not cause pressure drops in the gas piping system, the excess flow valve must also be sized and installed in accordance with the manufacturer's instructions.

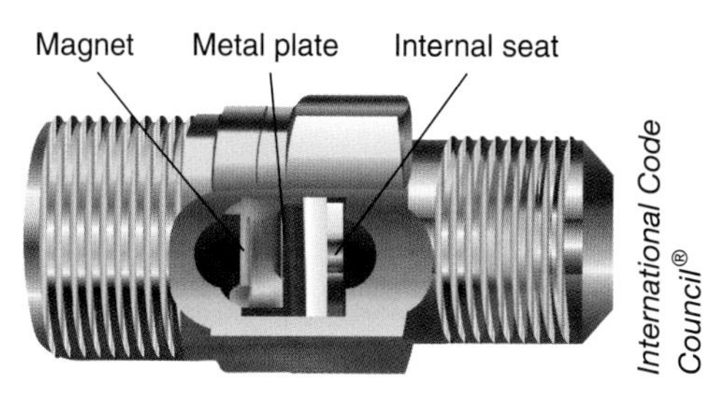

Listed excessive flow valve

202, 410.5

Flashback Arrestor Check Valve

CHANGE TYPE: Addition

CHANGE SUMMARY: A combination flashback arrestor and backflow check valve is now required on any fuel gas system used with oxygen in any hot work operation.

2012 CODE:

SECTION 202
DEFINITIONS

Flashback Arrestor Check Valve. A device that will prevent the backflow of one gas into the supply system of another gas, and prevent the passage of flame into the gas supply system.

410.5 Flashback Arrestor Check Valve. Where fuel gas is used with oxygen in any hot work operation, a listed protective device that serves as a combination flashback arrestor and backflow check valve shall be installed at an approved location on both the fuel gas and oxygen supply lines. Where the pressure of the piped fuel gas supply is insufficient to ensure such safe operation, approved equipment shall be installed between the gas meter and the appliance that increases pressure to the level required for such safe operation.

CHANGE SIGNIFICANCE: To prevent flashbacks from occurring and prevent oxygen from getting into the gas supply system, fuel gas systems that are used with oxygen are now required to have a flashback arrestor and a backflow check valve on the gas supply line and the oxygen supply line.

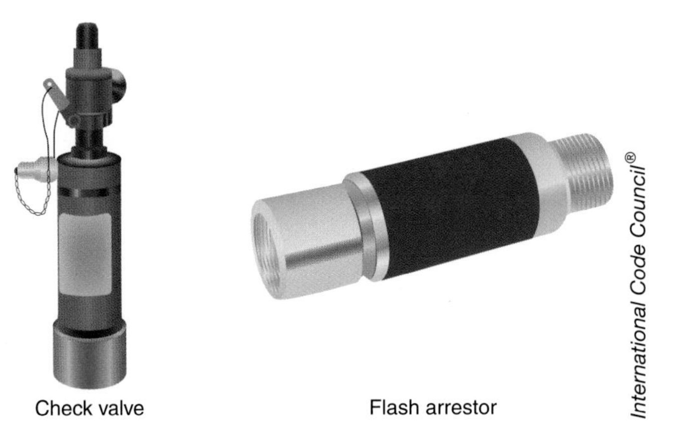

Check valve and flash arrestor

618.4

Prohibited Sources

CHANGE TYPE: Modification

CHANGE SUMMARY: Return air may be taken from a garage provided with a dedicated forced-air system.

2012 CODE: **~~618.5~~ <u>618.4</u> Prohibited Sources.** ~~Outside~~ <u>Outdoor</u> or return air for forced-air heating <u>and cooling</u> systems shall not be taken from the following locations:

1. Closer than 10 feet (3048 mm) from an appliance vent outlet, a vent opening from a plumbing drainage system or the discharge outlet of an exhaust fan, unless the outlet is 3 feet (914 mm) above the outside air inlet.
2. Where there is the presence of objectionable odors, fumes or flammable vapors; or where located less than 10 feet (3048 mm) above the surface of any abutting public way or driveway; or where located at grade level by a sidewalk, street, alley or driveway.
3. A hazardous or insanitary location or a refrigeration machinery room as defined in the *International Mechanical Code*.
4. A room or space, the volume of which is less than 25 percent of the entire volume served by such system. Where connected by a permanent opening having an area sized in accordance with Section 618.2, adjoining rooms or spaces shall be considered as a single room or space for the purpose of determining the volume of such rooms or spaces.

   Exception: The minimum volume requirement shall not apply where the amount of return air taken from a room or space is less than or equal to the amount of supply air delivered to such room or space.

618.4 continues

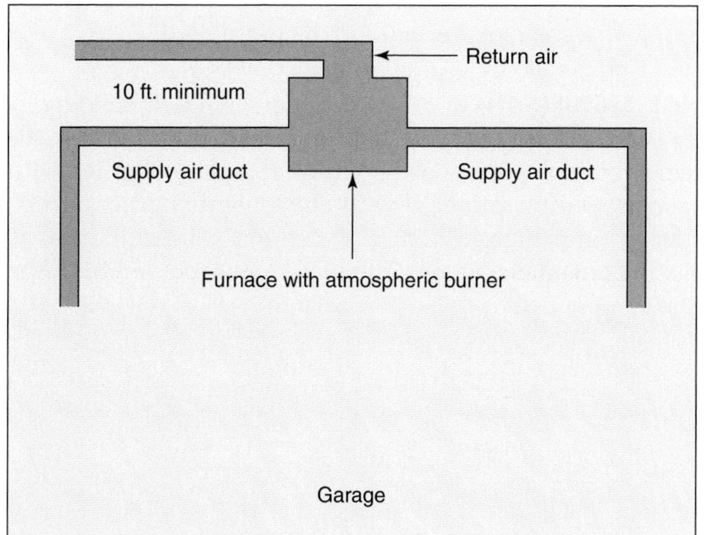

Furnace serving a garage

618.4 continued

5. A room or space containing an appliance where such a room or space serves as the sole source of return air.

Exception: This shall not apply where:

1. The appliance is a direct-vent appliance or an appliance not requiring a vent in accordance with Section 501.8.
2. The room or space complies with the following requirements:
   - 2.1. The return air shall be taken from a room or space having a volume exceeding 1 cubic foot for each 10 Btu/h (9.6L/W) of combined input rating of all fuel-burning appliances therein.
   - 2.2. The volume of supply air discharged back into the same space shall be approximately equal to the volume of return air taken from the space.
   - 2.3. Return-air inlets shall not be located within 10 feet (3048 mm) of a draft hood in the same room or space or the combustion chamber of any atmospheric burner appliance ~~firebox or draft hood~~ in the same room or space.
3. Rooms or spaces containing solid fuel-burning appliances, provided that return-air inlets are located not less than 10 feet (3048 mm) from the firebox of such appliances.

6. A closet, bathroom, toilet room, kitchen, garage, ~~mechanical room,~~ boiler room, furnace room or unconditioned attic.

Exceptions:

1. Where return air intakes are located not less than 10 feet (3048 mm) from cooking appliances and serve only the kitchen area, taking return air from a kitchen area shall not be prohibited.
2. Dedicated forced air systems serving only a garage shall not be prohibited from obtaining return air from the garage.

7. A crawl space by means of direct connection to the return side of a forced-air system. Transfer openings in the crawl space enclosure shall not be prohibited.

CHANGE SIGNIFICANCE: A dedicated forced-air heating system that is only serving a garage may take return air from the garage where the furnace is located, provided the return air inlet complies with item 5 in this section. Only appliances with atmospheric burners need to comply with the requirements in item 5, Exception 2.3. Appliances with closed combustion chambers are not intended to be included in this exception.

Index

R

S

Don't Miss Out On Valuable ICC Membership Benefits. Join ICC Today!

Join the largest and most respected building code and safety organization. As an official member of the International Code Council®, these great ICC® benefits are at your fingertips.

EXCLUSIVE MEMBER DISCOUNTS

ICC members enjoy exclusive discounts on codes, technical publications, seminars, plan reviews, educational materials, videos, and other products and services.

TECHNICAL SUPPORT

ICC members get expert code support services, opinions, and technical assistance from experienced engineers and architects, backed by the world's leading repository of code publications.

FREE CODE—LATEST EDITION

Most new individual members receive a free code from the latest edition of the International Codes®. New corporate and governmental members receive one set of major International Codes (Building, Residential, Fire, Fuel Gas, Mechanical, Plumbing, Private Sewage Disposal).

FREE CODE MONOGRAPHS

Code monographs and other materials on proposed International Code revisions are provided free to ICC members upon request.

PROFESSIONAL DEVELOPMENT

Receive Member Discounts for on-site training, institutes, symposiums, audio virtual seminars, and on-line training! ICC delivers educational programs that enable members to transition to the I-Codes®, interpret and enforce codes, perform plan reviews, design and build safe structures, and perform administrative functions more effectively and with greater efficiency. Members also enjoy special educational offerings that provide a forum to learn about and discuss current and emerging issues that affect the building industry.

ENHANCE YOUR CAREER

ICC keeps you current on the latest building codes, methods, and materials. Our conferences, job postings, and educational programs can also help you advance your career.

CODE NEWS

ICC members have the inside track for code news and industry updates via e-mails, newsletters, conferences, chapter meetings, networking, and the ICC website (www.iccsafe.org). Obtain code opinions, reports, adoption updates, and more. Without exception, ICC is your number one source for the very latest code and safety standards information.

MEMBER RECOGNITION

Improve your standing and prestige among your peers. ICC member cards, wall certificates, and logo decals identify your commitment to the community and to the safety of people worldwide.

ICC NETWORKING

Take advantage of exciting new opportunities to network with colleagues, future employers, potential business partners, industry experts, and more than 50,000 ICC members. ICC also has over 300 chapters across North America and around the globe to help you stay informed on local events, to consult with other professionals, and to enhance your reputation in the local community.

JOIN NOW! 1-888-422-7233, x33804 | www.iccsafe.org/membership

People Helping People Build a Safer World™

09-01530

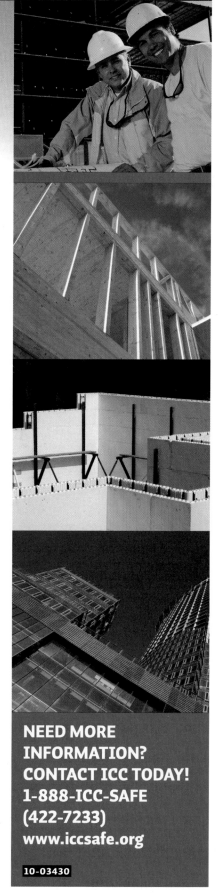